D0594167

NUMBER

MIDHAT GAZALÉ

NUMBER

From Ahmes to Cantor

PRINCETON UNIVERSITY PRESS

PRINCETON, NEW JERSEY

Library of Congress Cataloging-in-Publication Data
Gazalé, Midhat J., 1929–
Number : from Ahmes to Cantor / Midhat Gazalé.
p.   cm.
Includes bibliographical references and index.
ISBN 0-691-00515-X (cloth : alk. paper)
1. Numeration—History.   2. Number theory—History.   I. Title.
QA141.G39   1999
513—dc21   99-36677

This book has been composed in Berkeley
The paper used in this publication meets the minimum requirements of
ANSI/NISO Z39.48-1992 (R1997) (*Permanence of Paper*)

http://pup.princeton.edu

Printed in the United States of America

1 3 5 7 9 10 8 6 4 2

To Fabio, Clara and Helena

☥ **Contents** ☥

☥ **Preface** ☥

During my student years at the Faculty of Engineering, I used to ask myself a number of fundamental questions about number, whose answers were not being provided within the confines of an engineering curriculum. How did the decimal notation come about? Are the decimal and binary systems the only legitimate ones? Are there nonuniform base number systems? Exactly what are the irrationals? Why is their representation not periodic? What is a transcendental number? What is a real number? What is the famous continuum? What is meant by infinity? And so on. Following my doctoral work, I resolved that I would someday write a book on these subjects, aimed at an audience possessing a reasonably good understanding of college mathematics and inquisitive minds, in the hope that it would relieve them from the kind of frustration that I was experiencing.

It is said that the great Gauss imposed upon himself the systematic occultation of anything that might hint at the laboriousness of his discoveries. Nothing discourages creativity by gifted students as much as the neatly packaged, coldly presented mathematical objects that they are given to accept and learn. Another frustration therefore stemmed from the manner in which mathematical truths were being presented. They looked as though they had descended upon a select group of mathematical prophets in an effortless manner, transcending time and place. I therefore decided that, as often as possible, I would replace the mathematical objects (which, depending on the reader's philosophical outlook, he or she may regard as discoveries or constructions) within their historical context, showing, wherever possible, the effort that led to their elaboration.

Ce qui se conçoit bien s'énonce clairement! That which is well conceived is clearly stated. A third frustration stemmed from the difficult language deliberately used by some mathematicians to describe even the simplest of objects (particularly in France, in the wake of the Bourbaki episode).[1] I have therefore adopted a step-by-step approach,

---

[1] In 1939, a group of mathematicians, most of them French, published the first of more than twenty volumes, using the pseudonym Nicolas Bourbaki, who allegedly taught at the University of Nancago (a contraction of Nancy and Chicago). Their common object was the reformulation of the entire mathematical edifice using the same

and used the simplest possible language, perhaps to the dismay of seasoned mathematicians, and at the expense of absolute thoroughness.

In his *Solutions analytiques de quelques problèmes sur les pyramides triangulaires*, published in 1775, the great French mathematician Joseph Louis Lagrange boasted that his solutions of geometry problems were purely analytic and could be understood without figures. Strange as that may seem, his geometry book did not include a single figure.[2] Perhaps one rare illustration here and there, sparsely dispensed, but that is all. Following my natural inclination, I decided to resort to geometrical and physical metaphors as often as afforded by the subject at hand.

Upon finally sitting down to write a book, I discovered that my mind was wandering from one subject to another. The result was *Gnomon: From Pharaohs to Fractals*,[3] which deals with self-similarity, and this book, which deals with number and number representation.

This book opens with a study of the fundamental notions underlying the acquisition and recording of number, such as matching, or *cardination*, and counting, or *ordination*. There follows a mathematical-historical account of three archaic number systems, namely, the Egyptian *additive* system and the Mesopotamian and Mayan *positional* systems. The two presently used systems, the Hindu-Arabic decimal system and binary numeration are then examined.

A detailed examination is then offered of *mixed* and *uniform* positional systems. The object is to dispel the notion that the family to which decimal and binary number systems belong, that of *uniform* positional systems, might be the only possible one. An infinity of nonuniform bases is conceivable, some of which offer interesting applications, such as the analysis of certain classes of fractals, which were examined in *Gnomon*, as well as tensors, perhaps the subject of a forthcoming book.

Positional numeration having been established on firm grounds, the next chapter examines *divisibility*, as it relates to positional numeration. The questions it purports to answer are the following:

---

symbolism throughout. These mathematicians included Dieudonné, Cartan, Eilenberg, and Weil, among others.

[2]It is Lagrange who observed, in a 1781 letter to his mentor, Jean le Rond d'Alembert, that no further progress could be expected in mathematics, which had reached its limit!

[3]Midhat J. Gazalé, *Gnomon: From Pharaohs to Fractals* (Princeton, N.J.: Princeton University Press, 1999).

Why is the positional representation of irrational numbers not periodic? How can the period of rational numbers be predicted? To that effect, Euler's theorem is explained, and a particular pedagogical effort is devoted to presenting somewhat arduous number-theoretic results in an abundantly illustrated and hopefully pleasant form. I then present a generalization of Euler's theorem of my own, and use it to explain the periodic character of rational number representations. The chapter closes with a study of cyclic numbers and strings of zeros and ones, as well as Mersenne primes.

The next chapter dwells on the notion of *real numbers*, and a broad-brush presentation is given of the Frege-Russell and Peano definitions, as well as Brouwer's intuitionistic view. There follows a study of the integral domain, and an axiomatic approach to the rational number field.

In order to prepare the reader for the study of irrational numbers, the theorem of Pythagoras is first examined, because it led to the famous proof of the irrationality of $\sqrt{2}$, usually attributed to Euclid. The Ladder of Theodorus of Cyrene is presented as an illustration of a particularly relevant Diophantine equation (Pell's equation). A simple presentation is then given of Dedekind's *Schnitt*, as well as of the historic definition of irrationals given by Eudoxus.

The next chapter introduces the reader to the neglected field of *continued fractions*, as preparation for the notion to be next introduced, that of cleavages.

The chapter on *cleavages* constitutes an original contribution that graphically illustrates the axiom of Eudoxus, as well as Dedekind's *Schnitt*. It sheds some light on the mysterious nature of irrational numbers. En route, some relatively unknown properties of fractions are examined. These are the fractions generated by the little-known but fascinating *Stern-Brocot tree*.

The final chapter dwells on the difficult subject of *infinity*. Following an elementary study of convergence, some paradoxes of infinity are examined: Hilbert's Hotel, Zeno's paradoxes, and others. There follows a discussion of the exhaustion process, and an exposition of the debate between the advocates of *actual infinity* (Cantor), and the Aristotelians. Cantor's theory is explained in simple terms, by means of geometric and other metaphors.

I cannot close this brief preface without paying a special tribute to those individuals who unknowingly influenced me the most. First and foremost is Martin Gardner, whose acumen, wit, and unselfish kindness have inspired so many of the ideas presented in this and a previous book.[4] How can one thank Gardner without also saluting Professor Donald E. Knuth of Stanford University who is, in Gardner's own words, "extraordinary mathematician, computer scientist, writer, musician, humorist, recreational math buff, and much more." I personally treasure his *Fundamental Algorithms*,[5] as well as his *Concrete Mathematics*,[6] which is co-authored by Ronald Graham of Bell Laboratories and Oren Patashnik of the Center for Communications Research, and is appropriately dedicated to Leonhard Euler. Tobias Dantzig's *Number, the Language of Science* was a true revelation.[7] Indeed, as Einstein himself said, "This is beyond doubt the most interesting book on the evolution of mathematics which has ever fallen into my hands." In D'Arcy Thompson's *On Growth and Form*, amidst an extraordinary display of scholarship and humility, let alone its rare literary form, I first encountered the word *gnomon*, which became the theme of my first book. Ian Stewart is continuing the Gardner tradition with his own distinctive talent, and, like his predecessor, has the rare gift of making the most difficult subject look simple. I have often used John Horton Conway's Game of Life to illustrate my lectures on Von Neumann's cellular automata, and I am remiss for not having mentioned his surreal numbers in this book. I recommend reading his *Book of Numbers* (co-authored by Richard K. Guy),[8] Eli Maor's *To Infinity and Beyond* and *e, The Story of a Number*,[9] and George Ifrah's *Histoire universelle des chiffres*,[10] his lifetime truly encyclopedic magnum opus. I also wish to thank Professor Fathy Saleh of Cairo University for

[4]Martin Gardner's contributions are too numerous to be quoted here. I believe I have read every one of his columns in *Scientific American*, as well as every book he ever published (some more than once).

[5]Donald E. Knuth, *Fundamental Algorithms* (Reading, Mass.: Addison-Wesley, 1969).

[6]Ronald L. Graham, Donald E. Knuth, and Oren Patashnik, *Concrete Mathematics* (Reading, Mass.: Addison-Wesley, 1994).

[7]Tobias Dantzig, *Number, the Language of Science*, 4th ed. (Garden City, N.Y.: Doubleday, 1954).

[8]John Horton Conway and Richard K. Guy, *The Book of Numbers* (New York: Copernicus, 1995).

[9]Eli Maor, *To Infinity and Beyond* (Princeton, N.J.: Princeton University Press, 1991); id., *e, The Story of a Number* (Princeton, N.J.: Princeton University Press, 1994).

[10]George Ifrah, *Histoire Universelle des Chiffres* (Paris: Editions Robert Laffont S.A., 1994).

introducing me to ancient Egyptian mathematics, as well as Trevor
Lipscombe of Princeton University Press, without whom this book
might have never happened.

I cannot think of a more appropriate way to close this brief
"mathematician's apology" than to quote the greatest of all Renaissance men:

> My concern is to find cases and inventions gathering them as they
> occur to me . . . therefore, you will not wonder nor will you laugh
> at me, Reader, if here I make such great jumps from one subject to
> another. (Leonardo da Vinci, *The Leicester Codex*)

NUMBER

A mathematician, like a painter or poet, is a maker of
patterns . . . . The mathematician's patterns, like the painter's,
or the poet's, must be beautiful; the ideas, like the colors
or the words, must fit together in a harmonious way.
(G. W. Hardy)

It has often been said that number is as old as private property, and
it is believed that the recording of numbers preceded that of language.
While it is true that geometry, literally meaning "earth measurement,"
and the recording of other measurable quantities were necessary for the
emergence of economies, the establishment of authority and might, and
the administration of justice, to infer that early mathematicians were
mere accountants and land surveyors is to deny the creative genius
of man, his fundamental fascination with the mysteries that surround
him, his intuition of the necessary and the contingent, his aspiration to
a coherent interpretation of the universe, and perhaps above all, his
innate sense of beauty.

Could it be that the man or woman who painted the walls in the
cave of Altamira or that of Lascaux was moved more by artistic or
religious awe than by some down-to-earth purpose? Could it be that
the contemporary who carved tally marks on the bones of a wolf was
performing another kind of metaphor, akin to that of his or her fellow
artist? Could it be that both men or women felt a movement of the spirit
more than the sense of having properly recorded the number of bison
in a herd? Are not the geometric designs found on primitive artifacts
early abstractions of natural forms encountered in the wilderness, perhaps endowed with magical properties, or simply with a rhythm whose
resonance with some attribute of the spirit elicited joy and fulfillment?

Says Carl B. Boyer, "The concern of prehistoric man for spatial
designs and relationships may have stemmed from his aesthetic feel-
ing and the enjoyment of beauty of form, motives that often actuate the
mathematician of today."[1] Is it possible that Ahmes, the Egyptian scribe,

[1]Carl B. Boyer, *A History of Mathematics*, 2d ed., rev. by Uta C. Merzbach
(New York: Wiley, 1989).

when he sat down some thirty-eight centuries ago, to copy a long and intricate mathematical papyrus, was touched by a sense of harmony, or perhaps eternity, rather than one of merely doing his job, as we would unimaginatively say today?

> As for the learned scribes . . . they did not make for themselves pyramids of copper with tombstones of iron. They were unable to leave an heir in the form of children who would pronounce their name, but they made for themselves an heir of the writing and instruction they hade made. (From a hieroglyphic inscription on an ostracon found in Deir el Medina)

Was some obscure Indian genius, around 900 B.C., driven into inventing the zero by some "market pull," or "technology push," the two panaceas of this day, so dear to business schools, and the foundations of the education that our children receive?

Unreasoned fear in the face of the unknown, and hopes of a lasting betterment of the human condition, are more surely at the root of science than the concomitant earnest desire to keep a record of earthly possessions. Astrology gave birth to astronomy, alchemy to chemistry, embalming to surgery, theology to philosophy, mythology to science—and Jean-Jacques Rousseau cynically remarked that if geometry was born of avarice, astronomy was born of superstition.

Number theory is the child of number superstition and mysticism. Through the ages, mysterious powers were attributed to number, sometimes reaching unexpected heights with the numerology of the ancient Greeks and the innumerable forms of modern-day number superstition.

With these sciences, as they were thought to be, Gnostics calculated the number-name of God, and Peter Bungus, a Catholic theologian, inspired by an unrelenting hatred of Protestants, devoted no less than seven hundred pages, and a lifetime, to the futile task of proving that the gematric value of Martin Luther's name, 666, was the same as that calculated by St. John the Divine for the Antichrist. Quoting from Revelation (13:18), "Let him that hath understanding count the number of the Beast: for it is the number of a man; and that number is Six hundred threescore and six." In 1593, the Scottish mathematician John Napier, better known for the invention of logarithms than for his countless eccentricities and his opposition to papacy, published his popular *A plaine Discovery of the revelation of Saint John*, in which he proved that the pope was the Antichrist. The mysterious number 666 had also been calculated by another inspired numerologist to be that of Emperor Nero.

Not only were mystical and religious virtues bestowed upon number by members of the secret sect founded by Pythagoras in the southern Italian town of Croton, but number worship was one of the cornerstones of their philosophy. A prayer to the holy Tetraktys was addressed by the Pythagoreans to a pile of ten circles resembling a heap of cannonballs. They chanted, "O holy holy Tetraktys, thou that containest the root and the source of the eternally flowing creation! For the Divine Number begins with the profound, pure Unity until it comes to the holy Four; then it begets ... the never tiring holy Ten, the keyholder of all!" To the Pythagoreans, One was the mind, Two was opinion, allowing for wavering and confrontation, and Three was reconciliation and wholeness, Four was justice, perhaps the precursor, according to a facetious Rudy Rucker, of our modern-day "square deals".[2] Five was marriage, as Two not only stood for wavering but also for woman, and Three, in addition to wholeness, also stood for man. (Obviously, in the minds of the Greeks, long before the French, *souvent femme varie*.[3]) So strong was their belief in the magical virtues of whole numbers, that they refused to acknowledge the existence of the αλογον, the "unutterable" irrational numbers, and it is even said that poor Hippasus was thrown overboard in the course of an outing at sea, for fear that he would reveal their existence.

Herodotus described Egyptian geometry as the art of land measurement, or rope-stretching. In his mind, that strictly utilitarian craft provided the foundation upon which, beginning with Thales, the Greeks built their abstract geometrical edifice. The emergence of science, in his opinion, thus stemmed from necessity.

Such was not the view of Aristotle, who was inclined to trace the emergence of science to more contingent human drives, stemming from intuition and creative genius. In his opinion, that genius could best express itself if scientists were relieved of the burden of ensuring their own subsistence and that of their next of kin, authorizing them to let their minds wander freely, without fear of tomorrow. He noted that mathematics was developed in Egypt by an elite of priests who enjoyed the leisure of studying under the protection of the pharaoh.

Opposed as these views may be, that of the necessary and that of the contingent, the truth probably lies somewhere in between, as it invariably does, and we do not have sufficient documentation to form definitive opinions on the matter.

[2]Rudy Rucker, *Infinity and the Mind* (New York: Bantam, 1983).
[3]Many a time and oft does women waver (vary).

Perhaps one invariant of human activity is that intellectual elites, relatively shielded from material constraints, but nonetheless immersed in the realities of their day (and who was more immersed in the bloody war between Greeks and Romans, to which he eventually fell victim, than the great Archimedes himself?) are authorized to dwell on abstract objects, not accessible to the ordinary citizen, and formulate the paradigms of their time, providing their society with a necessary contemporary vision of the world. That vision would then be translated into progress or devastation by others, inclined by their different nature to build, conquer, and rule.

Carl Friederich Gauss, one of the greatest mathematicians of all time, and founder of number theory as we know it today, wrote the dedication of his monumental *Disquisitiones arithmeticae* to his protector, Duke Ferdinand. In it, he praises the duke for having supported endeavours "which appear most abstract and with less application to ordinary usefulness, because in the depth of your wisdom, able to profit by all which tends to the happiness and prosperity of society, you have felt the intimate and necessary liaison which unites all sciences."

Thales, the founder of Greek geometry, who also successfully predicted the solar eclipse of 585 B.C., eventually retired from commerce and trade, and devoted the rest of his life to astronomy. As he was gazing at the stars one evening, he fell into a ditch, whereupon the old lady attending him exclaimed, "How can you know what is happening in the heavens when you do not see what is at your feet?"

Anaxagoras, who lived around 500 B.C., is said to have forsaken his considerable wealth in order to fully devote himself to what he felt was the object of being born: the contemplation of the sun, moon, and heavens. Because he declared that the sun was a giant red-hot stone, and moonshine was just the sun's reflected light, he was sentenced to be exiled. During his exile, he worked on the problem of squaring the circle, which was to baffle mathematicians for centuries to come.

Stobaeus tells the story of a student of Euclid's who, upon being taught the first theorem, asked the Master what he would gain by learning that sort of thing, whereupon Euclid ordered his slave to give the student three coins, "since he needs make profit of what he learns!" Aristoxemus, in "Elements of Harmony," said of the people who flocked into Athens to listen to the teachings of Plato, "but when they found that Plato's arguments were of mathematics and number and geometry and astronomy, and that at the end he declared the One to be the Good, ... they were altogether taken by surprise. The result was that some of them scoffed at the thing, while others found great fault in it."

Twenty-six centuries later, busloads of tourists visiting St. Patrick's
in New York hear of the cathedral's height, compared to that of the
Empire State Building, and of its volume in cubic feet, compared to St.
Peter of the Vatican, perhaps oblivious of the spiritual content of such
impressive volumes. But their appetites are not satiated until they are
told how much it cost to build the edifice, the common denominator
of all values. As George Bernanos once said in a singularly uncharita-
ble mood, "Machine Age is the civilization of Quantity, opposed to that
of Quality. Fools dominate by their number—They are number." Our
Machine Age has thus begotten a new mystique of number, not reserved
to a Pythagorean brotherhood or a handful of gematricians, but unfortunately pervading the entire Western world. A mathematician's toil is
relevant to the business community inasmuch as it produces tangible
marketable goods, as attested by this candid statement in *Business Week*
magazine: "As the world goes digital, it seems, the musings of number
theorists will only become more relevant."[4] As Bertrand Russell wrote,

Cartoon by Serguei. Courtesy of the artist.

"In the streets of a modern city, the night sky is invisible; in rural areas,
we move in cars with bright headlights. We have blotted out the heavens, and only a few scientists remain aware of the stars and planets,
meteorites and comets."

[4]"Suddenly, Number Theory Makes Sense to Industry," *Business Week*, June 20,
1994.

CHAPTER 1

The Genesis of Number Systems

FOUNDATIONS

To understand a science
is to know its history.
(Auguste Comte)

Once upon a time, a Mesopotamian sheepherder decided to keep a record of the sheep in his herd. He forced his sheep to cross a narrow passage, one at a time, as he dropped a pebble in a jar with each crossing. In so doing, he performed a so-called one-to-one correspondence, or mapping, between the sheep in the herd and the pebbles in the jar. At some later date, the sheepherder repeated the process in reverse, removing a pebble from the jar with each crossing, and in so doing ascertained that the quantity of sheep had not varied, assuming that the jar had not been tampered with, and could be regarded as an immutable reference, a permanent record. As he became wiser, he asked the village potter to provide him with a jar whose clay was soft enough to receive the imprint of objects pressed against it. Before he dropped each pebble in the jar, he first pressed it against its wall, where it left a visible imprint (Figure 1.1). At the end of the tally, the jar was sealed. As the sheepherder became wiser still, he realised that he no longer needed the pebbles, as the markings themselves adequately served his purpose. Without the pebbles, the jar did not need to be a jar after all. Instead, he took a flat clay tablet, and pressed the tip of a reed against it, leaving a notch with each crossing.

That is probably how early Sumerians embarked upon the long voyage of number recording, more than five thousand years ago. Doing away with the pebbles took centuries—perhaps until the Sumerians eventually felt comfortable enough with the soundness of their recording scheme, as well as with the population's "number literacy," to no longer require the "redundancy check" afforded by the pebbles.

Figure 1.1. Mesopotamian clay tallying jar and calculi, ca. 3300 B.C.
Louvre Museum. Courtesy Réunion des Musées Nationaux, Paris.
Photo by R. G. Ojeda.

Matching

The scheme just described constituted the most elementary of
all *additive* number systems: Only the number of notches mattered,
not the position of any particular notch within the array. Despite the
sequential character of the processes described, the establishment of
such one-to-one correspondence, or *matching*, is entirely distinct from
the act of counting, which assigns a successor to every number. In the
matching scheme, each pebble is dropped independently of previous
events and of the quantity of pebbles accrued in the jar. A more appropriate metaphor for the act of matching might consist of unleashing
the herd of sheep in an open field planted with cabbage, and allowing
each sheep to eat no more than one head, while making sure that it
eats at least one. In that case, matching is performed *le tout ensemble*,
or "in parallel," to use the language of modern-day computer designers. At the end of the feast, the quantity of missing cabbage matches
that of the sheep.

Experiments with birds have shown that they are endowed with
the capacity of assessing quantities all at once, albeit within very

narrow limits, beyond which the objects become too numerous for the birds to match against some learned pattern. Crows apparently cannot discern quantities in excess of four. It also seems that humans can at once discern a single object, a pair, a threesome, and a grouping of four objects. It is probably difficult to construct experiments aimed at measuring the limits of that *cardinal* sense, as educated humans invariably and irresistibly fall back on their learned *ordinal* capability, as they resort to counting.

Naming

> In every human tongue there are rules, and each
> term at placitum signifies one object according
> to an unchanging rule, because man cannot in one
> instant call a dog a dog, and a cat in another.
> (Umberto Eco, *The Name of the Rose*)

Correspondence, or matching, consists of pairing objects or groups of objects, according to some shared attribute. That attribute may be color, size, shape, etc. It may also be quantity, as in the case of sheep and pebbles. It is a common feature of every language that the name of one of the members of the matched pair should often become the name of the attribute itself. *Azur* in French signifies sky as well as light blue. In Arabic, *aich* represents both bread and life, and *nafas* is the root of the words for breath and individual.

Nomen est numen says the Latin adage, meaning that to name is to know. Man's earliest remarkable achievement in the realm of reckoning was without any doubt the *naming* of numbers, or the assignment of vocabulary to quantity. Not only did *verbal* communication become possible between individuals, but the emergence of counting and calculating could take place. (To this day, when we perform an addition or multiplication, we resort to our verbal competence everytime we recall that two and three is five, or four times five is twenty.) Early humans sought names for quantities amongst their most immediate environment, their own bodies, confirming the observation by Marcel Mauss that "the human body is the first and most natural instrument of man." In many primitive languages, the number five is expressed by "hand," and the number ten by "two hands," or by "man." According to Tobias Dantzig, the Sanskrit *pantcha*, meaning "five," comes from the same root as the Persian *pentcha*, meaning "hand," and the Russian *piat* comes from *piast*, "the outstretched hand." The human hand constituted a rather sophisticated instrument, as the articulability of the fingers, allowing one to conceal or display a finger at will, enabled the establishment of one-to-one correspondence between animals in a herd and the fingers of one individual, or of several individuals jointly using their hands.

Counting

Did you bring me a man who cannot count on his fingers?
The Book of the Dead

One of the most fundamental processes ever invented, counting, which consists of enunciating number names in successive increments of one unit, was to become one of the cornerstones of all subsequent human progress. Counting may be second nature to the number literate, but it has to be learned, as attested by the difficulty we experience in counting backwards, a process that mobilizes our attention to a much larger degree. Recitation of numbers in succession was not known, and may still not be known, odd as that may seem, to many so-called primitive tribes, whereas our children take great pride in being able to count to ten, or one hundred—or, better still, in increments larger than one, such as two, ten, etc. The process of associating number words with number, and associating every number word with a successor, is the deep underpinning of number literacy, which radically differentiates us from other animal species. Recent studies, however, seem to also indicate that monkeys are capable of ordination, albeit within narrow limits. They can determine which of two numbers is the larger, and extrapolate their experience to substantially larger numbers.

According to studies conducted by Charles J. Brainerd and others, the faculty of ordination (establishing an asymmetrical relation of transitivity between three balls of increasing weight, or three sticks of increasing length) is more fundamental than that of cardination or matching. Experiments conducted with children seemed to reveal that "two normative studies showed an invariant sequence in the growth of

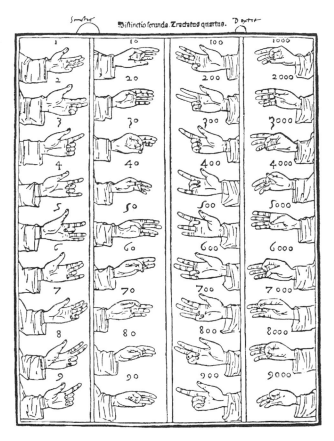

Figure 1.2. Counting with fingers. From Luca Pacioli's *Summa arithmetica.*

the children's concept of number. Ordination was the first to emerge, followed by natural number competence and then by cardination."[1]

In French, *compter*, "to count," and *conter*, "to recount" (a fable, a story, an event), are, along with their English equivalents, derived from the same Latin root, *computare*. Both verbs imply the recitation of items in a well-defined sequence, each following its predecessor, as with the letters of the alphabet. Children spend several hours of their early school days reciting two fundamental word sequences, that of the alphabet and that of integers, along with the addition and multiplication tables, until the "three R's" become second nature to everyone:

---

[1]Charles J. Brainerd, "The Origins of Number Concepts," *Scientific American,* April 1973, pp. 101–109.

"Reading, 'Riting, 'Rithmetic / Taught to the tune of a hickory stick." Other sequences that are also learned in childhood are the names of the days of the week and the months of the year, the names of the seasons, the musical notes, etc.

Says Tobias Dantzig, "Correspondence and succession, the two principles that permeate all mathematics—nay, all realms of exact thought—are woven into the very fabric of number systems."[2] For all we know, and that does not amount to much, they are perhaps woven into the very fabric of our brain hemispheres, a predominantly *cardinal* right brain, and a predominantly *ordinal* left brain, and whereas an innate number sense is often alluded to, one might conjecture that the more fundamental senses are that of correspondence, or pairing or matching, and that of sequencing, or ordering, *perhaps* with a specific mapping on the brain for each of the two senses. It seems that individuals suffering from aphasia, often caused by left brain impairment, lose the ability to count as well as that of recognizing words as successions of letters of the alphabet. They are nonetheless capable of global written-word recognition, as well as an overall appreciation of quantity: few, many, very many. On the other hand, individuals suffering from agnosia, which often accompanies right brain impairment, may be very adept at detecting spelling or grammatical errors in a text, but fail to comprehend the gist of it. They may read large numbers but fail to appreciate the bigness or smallness that those numbers convey). That dichotomy is illustrated by the story entitled "The President's Speech," in Oliver Sacks's beautiful *"The Man Who Mistook His Wife for a Hat.*[3] To quote Wendy Heller, "The right hemisphere, it seemed, is skilled at grasping a general form—the gestalt—of an object. The left hemisphere is more successful at apprehending objects that can readily be analyzed into simpler features, easy to count or name: corners, edges, indentations."[4] "So complete is the intellectual merger, in everyday life, of the ordinal and cardinal that, simple as these notions may be, it is quite difficult to make the average educated person aware that they actually belong to two distinct processes. The most elementary of arithmetic operations draw upon a large array of faculties, some innate

---

[2] Tobias Dantzig, *Number, the Language of Science*, 4th ed. (Garden City, N.Y.: Doubleday, 1954), p. 9.

[3] Oliver Sacks, *The Man Who Mistook His Wife for a Hat* (New York: Summit Books, 1985).

[4] Wendy Heller, "Of One Mind," *The Sciences*, May–June 1990, pp. 38–44.

and some acquired, some verbal and some visual, not to mention other, more mysterious intuitive faculties that have yet to be understood.

Grouping

When Buddha reached manhood, he courted the beautiful Princess Gopa, and was submitted to a number of tests at the hands of the great mathematician Arjuna, who challenged him to name the number of atoms in a Yoyama, or mile. Buddha's answer begins as follows:

> Seven atoms make one very minute particle
> Seven very minute particles make one minute particle
> Seven minute particles make one that the wind will still carry
> Seven such particles make one rabbit track
> Seven rabbit tracks make one ram's track
> Seven ram's tracks make one ox track
> Seven ox tracks make one poppy seed
> . . . .

Perhaps that ancient Indian tale inspired this riddle in the form of an English children's rhyme.

> As I was going to Saint Ives,
> I met a man with seven wives.
> Ev'ry wife had seven sacks
> Ev'ry sack had seven cats
> Ev'ry cat had seven kits.
> Kits, cats, sacks and wives,
> How many were going to Saint Ives?

It is worth observing that problem 79 in the Rhind Papyrus also deals with groupings of seven objects: houses, cats, mice, ears of spelt, hekats of grain.

As the number of physical objects to be recorded becomes large, individual number names become too numerous to be remembered, and the number of imprints on a clay tablet, or indents on a tally bone become unwieldy. Since the dawn of civilization, man has thus endeavoured to group individual objects, and to assign names as well as written symbols to the groupings themselves. The reader may begin to imagine the gigantic undertaking that took centuries in unfolding, leading to the conceptual stroke of genius referred to as multiplicative

was a Scottish antiquary who suffered from tuberculosis and sojourned in Luxor in 1858, where the warm, dry climate attracted wealthy European tourists. There, he purchased the papyrus that now bears his name and was acquired by the British Museum following his death. Other portions of the papyrus, which had been sold to the American collector Edwin Smith, later turned up at the New York Historical Society. The Rhind Papyrus constitutes a handbook of practical everyday mathematics, with some eighty-seven problems aimed at teaching by example the art of fractions, the solution of elementary equations, the properties of progressions, and the mensuration of areas and volumes.

Smith, who lived in Luxor, also purchased the two most famous medical papyrii, today known as the Smith Papyrus and the Ebers Papyrus. They came from a clandestine find in the Ramses II necropolis. In those days, tomb plunderers abounded, and sold their loot to shrewd receivers. One of the manuscripts was later sold by Smith to the German Egyptologist Georg Ebers, and is named after him.

A few years earlier, in 1842, Champollion's *Dictionnaire Egyptien* had been posthumously published by the author's brother, following the publication of the first installment of his monumental opus, *La Grammaire Egyptienne*, on December 12, 1835, the very day that Champollion would have turned forty-five, had he not prematurely died in 1832. Jean-Francois Champollion (Plate 3) was born in Figeac, a small town in the southwest of France, and devoted his entire short life to a singularly focused goal, that of deciphering the Egyptian alphabet. He was extremely fluent in the Coptic language, which was spoken by the Egyptians at the dawn of Christianity, and at the age of sixteen, he delivered a paper before the Académie of Grenoble, in which he defended the thesis that the Coptic language was that of the ancient Egyptians. In 1822, he wrote his famous *Lettre à Monsieur Dacier*, which contained his first exposition of the phonetic character of hieroglyphs, following his initial encounter with the Rosetta stone.

The Rosetta stone (Figure 1.4) was unearthed in August 1799, during the Napoleonic expedition to Egypt, in the town of Rashid, or Rosetta, hence its name. Rashid lies on the western branch of the Nile delta, near Alexandria. Its discoverer, officer Pierre Bouchard, stumbled upon the stone in the course of the construction of Fort Julien, near Rosetta. The Hellenists who accompanied the expedition[6]

---

[6]A decision by the Directoire, dated March 16, 1798, gave birth to the Commission des Sciences et des Arts of the young republic's Armée d'Orient. Within a short

Figure 1.1. The Rosetta stone. Courtesy British Museum, London.

immediately understood that it was a copy of a decree handed down by Ptolemy V (196 B.C.), and hypothesized that the Greek text was the translation of the accompanying hieroglyphic text. Upon the defeat of the French army at the hands of the British, an English diplomat by the name of Hamilton discovered that the French general Menou was attempting to save the stone, and seized it on board a French ship in 1801. The Rosetta stone, which is kept in the British Museum, and was thoroughly cleaned in 1998, is really a slab of granite bearing three texts carved during the reign of the Ptolemies. It consists of an original Greek text and its Egyptian translation carved in both

---

three months, an impressive number of civilians were recruited, some of them in their twenties, and placed under the command of a general. Among them were engineers, technicians, mathematicians, chemists, physicians, naturalists, drafters, architects, orientalists, musicians, printers, and so on. These were led by outstanding scientists of the caliber of Monge, Berthollet, Dolomieu, and others.

the hieroglyphic and Coptic alphabets. It was deciphered in 1869 by Egyptologist Eisenlohr.

Ancient Egyptians evolved a clear, easy-to-use decimal number system, and invented a base 2 multiplication system akin to that used by modern-day computers. They published tables of unit fractions, which to this day puzzle mathematicians, and to which we shall later return. They knew how to square numbers, extract square roots, and determine the sum of arithmetic and geometric progressions. They were able to calculate the surfaces of triangles and trapezoids, as well as the volume of the frustrum of a square pyramid.

Egyptians used rudimentary trigonometric functions for the measurement of slope, notably the *seqt*, a precursor of today's cotangent. The corners of the Great Pyramid of Gizeh approach right angles within a fraction of a degree. So great is their confidence in the pyramid's alignment, that some modern-day astronomers have attempted to assess the pyramid's age based on the shift in its position relative to the polar star over the centuries.

Among other things, the Egyptians knew that the ratio of a circle's area to its circumference was equal to the ratio of the circumscribing square's area to its perimeter. Notwithstanding the judgment of many historians, that statement contained the germ of true geometry, not mere "rope-stretching." They equated the area of a circle whose diameter measures 9 units to that of a square whose side measures 8 units, thus positing that $\pi = (4/3)^4 \cong 3.1605$ (see Figure 7.3a).

Around the fifth century B.C., the Egyptian calendar year contained 365 days, 360 of which were evenly distributed among twelve months. The remaining *epagomenal* days were dedicated to five gods, and declared festive. Leap years were later accounted for, during the Roman Empire. In the judgment of Neugebauer, that was the "only intelligent calendar which existed in human history," and in that of J. W. S. Sewell, it was "the most scientific organisation of calendars which has yet been used by man."[7] Upon building the colossal temple of Abu Simbel, the Egyptians placed the statue of Ramses II at the far end of a deep sanctuary in such a manner that on the date of the king's accession to the throne, around October 20, and again in mid-February,

[7]Neugebauer and Sewell quoted in R. J. Gillings, *Mathematics in the Time of the Pharaohs* (Cambridge: MIT Press, 1972), pp. 236–240.

the rays of the rising sun fell exactly upon the statue and illuminated his face.

THE EGYPTIAN NUMBER SYSTEM

Transport yourself to ancient Egypt, and imagine that Ahmes the scribe is given a heap of amulets whose number he is required to record on a papyrus roll. The scribe's assistant, who was only taught to count up to 10, is given a supply of little bags, each of which bears the symbol ⌢, which represents the hobble of a water buffalo. The assistant is then instructed to obey the following simple rule: "Remove the amulets one at a time from the heap, and count them as they are being removed. Every time you reach the count of 10, drop the amulets in a bag, set the bag aside, and then repeat the process with the remaining amulets. When you have exhausted the heap, you may have some leftover amulets, not exceeding nine in number. Give these amulets to me." The assistant performs the task as instructed, and hands five amulets to his master, who writes down the number as five straight strokes (Figure 1.5a).

Figure 1.5a. Five amulets.

Ahmes now gives his assistant a new collection of larger bags, each bearing the symbol ℒ, which represents a coil of rope, and instructs him as follows: "Treat the hobble bags as if they were amulets and the coil bags as if they were hobble bags, and follow the previous procedure." The assistant performs the second counting cycle, hands six leftover hobble bags to his master, and awaits further instructions. The scribe extends his initial record as shown in Figure 1.5b.

Figure 1.5b. Sixty-five amulets.

He then provides his assistant with a collection of larger bags, with a little lotus flower, the symbol 𝈋, embroidered upon each of them.

The next instruction is the same as the previous one, with the symbols 𝍠 and 𝍪 respectively substituted for 𝍪 and ∩. At the end of the third cycle, the assistant hands his master four leftover coil bags, and the scribe extends his record as shown in Figure 1.5c.

Figure 1.5c. 465 amulets.

As the scribe observes that only two lotus bags are left uncounted, he completes his record as shown in Figure 1.5d. That inscription

Figure 1.5d. 2,465 amulets.

signifies that the number of amulets is 2,465. Obviously, more sophisticated counting methods were imagined by the Egyptians, not physically requiring the little bags. These are merely intended here as metaphors for groupings of objects in powers of 10.

Figure 1.5e. The number 2,123,013.

Figure 1.5e shows another example, involving a much larger number, and revealing the absence of zero. Figure 1.5f shows the icons for the first seven powers of 10. For another example of the Egyptian number icons, see Plate 4.

An imaginary educational pharaonic toy, aimed at teaching children how to write numbers, might consist of a collection of little wooden blocks with the number symbols engraved upon them. No more than nine identical blocks are required for any power of 10, and a total of $7 \times 9 = 63$ blocks suffices to represent any number between 1 and 9,999,999. Having constucted a given number's representation, a young student may thereupon shuffle the blocks to any

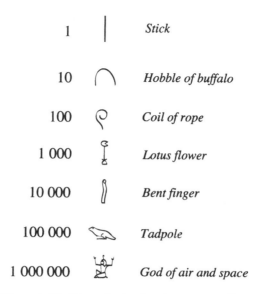

Figure 1.5f. Hieroglyphs for the powers of ten.

desired degree. The shuffled blocks would still represent the number, albeit in an unorthodox arrangement, and a patient tutor would be capable of restoring them to their correct positions without ambiguity. This example illustrates the essence of additive systems. The Egyptian numbering scheme epitomizes the decimal additive system, with a distinct symbol for every power of 10, where no more than nine symbols of any one kind need appear in any given number's representation.

Imagine a similar situation with a present-day child, equipped with a collection of decimal cubes, bearing the Arabic numerals. If, upon having constructed the number 2,465, the child were to shuffle the pieces, there would be no way of finding out what the intended number was, as there would be 4! = 24 equally probable guesses. Our decimal system, which is based on the principle of local value, is said to be positional, meaning that each symbol's contribution is equal to the value assigned to that symbol, multiplied by the power of 10 conferred by its localization, or position. In the pharaonic system, on the other hand, it is the symbol itself, not its position, that confers the corresponding power of 10.

The pharaonic example also illustrates the recursive, or iterative, character of counting. Had the instructions been given by Ahmes to a computer, they would have been executed in keeping with what modern-day computers do all the time, namely, repeat the same operations over and over, as "bag labels" are adjusted from one program step to the next.

The number system was not the Egyptians' only achievement in the realm of arithmetic. It went hand in hand with elaborate arithmetical procedures that remained in use beyond the Coptic and Greek eras, as late as the Byzantine Empire (around 540).

To this day, it is not entirely clear how the Egyptians performed addition and subtraction. According to some authors, they resorted to addition tables, though that is unlikely, as some trace of these tables would have surely been unearthed.

Multiplication techniques, on the other hand, are fully documented. They are based on what we would today refer to as the binary system. When the scribe needed to multiply two numbers, say, 11 times 9, he first had to decide which was the multiplier and which the multiplicand. If 9 were chosen to be the latter, the scribe would repeatedly multiply 9 by 2 and write down the resulting numbers in succession, as in the left panel of Figure 1.6a. Facing these, he would write down the successive powers of 2. He would then select those powers that added up to the multiplier (indicated by arrows in the figure), and add the corresponding duplicates in the right hand column, ignoring the remaining powers of 2 (indicated by round dots in the figure). In our example,

$$11 = 8 + 2 + 1; \text{ thus } 72 + 18 + 9 = 99.$$

	→ 1	9			→ 1	11
	→ 2	18			• 2	22
	• 4	36			• 4	44
	→ 8	72			→ 8	88
Total	11	99			9	99

Multiplicand 9 Multiplicand 11

Figure 1.6a. Multiplying 9 by 11.

The right panel of Figure 1.6a corresponds to multiplier 9 and multiplicand 11. Obviously, the scribe had to master the art of duplication, or duplation, as well as that of ascertaining which powers of 2 add up to a given number (multiplication literally means multiple duplications). Implicit in that method was the justified belief that any positive integer can be equated to the sum of a finite number of powers of 2.

$$
\begin{array}{rrl}
1 & 7 & \leftarrow \\
2 & 14 & \leftarrow \\
4 & 28 & \bullet \\
8 & 56 & \leftarrow \\
16 & 112 & \leftarrow \\
\hline
\end{array}
$$

Total   27   189

$$189 = 112 + 56 + 14 + 7$$
$$\text{thus quotient} = 16 + 8 + 2 + 1 = 27$$

Figure 1.6b. Dividing 7 into 189.

The process of division derived from that of multiplication. Dividing 7 into 189, as in Figure 1.6b, consisted of deciding which duplicates of 7 added up to 189, and then adding the corresponding powers of 2. In that example, the division of two numbers yields an integral quotient. As we shall later see, Egyptians were also quite familiar with fractional numbers.

In reality, Egyptians did not use arrows and dots; they simply struck out those lines corresponding to our dots. Imagine that the scribe had actually used arrows and dots, and, in a singular moment of enlightenment, decided to rotate the arrays by 90 degrees, declaring:

$$\uparrow \cdot \uparrow \uparrow = 11 \qquad \uparrow \cdot \cdot \uparrow = 9 \qquad \uparrow \uparrow \cdot \uparrow \uparrow = 27.$$

In the same breath, he would have exclaimed, "EUREKA!" or rather, the Egyptian equivalent thereof, "I have just invented binary numeration!" In all probability, that invention would not have been widely acclaimed. Convenient as it may be within today's highly computerized environment, it would have been found awkward for everyday use, and lacking in ergonomic convenience compared to the decimal system.

UNIT FRACTIONS

The Eye of Horus (Plate 5) or Wedjat Eye, is one of the best-known ancient Egyptian symbols. It may be found adorning jewelry and amulets, as well as hundreds of other artifacts. In Egyptian mythology, the eye of Horus was wrenched out by Seth, who was known for his cruelty. According to the Book of the Dead, the eye was made whole again by Toth, the Ibis-headed god, inventor of writing and mathematics. In the hieroglyphic alphabet, each portion of the eye represents the inverse of a power of 2 (see Plate 5). The Egyptians probably understood that the sum of the eye's parts fell short of 1. According to Gay Robins and Charles Shute, "If the Horus-Seth-Toth story had a mathematical connotation, it could be that the damaged Horus-eye was magically made whole by the restoration of the missing $1/64$."[8]

The Rhind Papyrus revealed that ancient Egyptians were extremely adept at representing fractional numbers as sums of unit fractions. They possessed extensive tables for the decomposition of a fraction into a sum of fractions with numerator 1, or unit fractions, no two of which were identical. For example, they would equate $11/14$ to $1/2 + 1/4 + 1/28$. The unit fraction $1/n$, or the reciprocal of n, was represented by the number n crowned by an open mouth, as in ⍁, which stands for $1/12$. One notable exception, the two-thirds fraction, was the essential ingredient of innumerable division recipes and enjoyed a special symbol, namely ⍕. The papyrus also contains elaborate tables for performing arithmetic with fractions, as in example R20 (Figure 1.7), which teaches how $(12 + 2/3) \times (7 + 7/8)$ is equated to $(90 + 9 + 1/2 + 1/4)$.

The following problem, literally translated, illustrates the manner in which the Egyptians solved most of the arithmetic problems they confronted in their daily lives:

A heap and its fifth, that makes twenty-one.

Figure 1.8 is an approximate transcription of that problem, as it was put to the scribe.

The scribe proceeded as follows: If I take a heap of 5 and add its fifth, that results in 6. The required sum, however, is 21. That number,

---

[8]Gay Robins and Charles Shute, *The Rhind Mathematical Papyrus, an Ancient Egyptian Text* (New York: Dover, 1987).

$$
\begin{array}{cc}
1 & 10 + 2 + \frac{2}{3} \\[1ex]
2 & 20 + 5 + \frac{1}{3} \\[1ex]
4 & 50 + \frac{2}{3} \\[1ex]
\frac{1}{2} & 6 + \frac{1}{3} \\[1ex]
\frac{1}{4} & 3 + \frac{1}{6} \\[1ex]
\frac{1}{8} & 1 + \frac{1}{2} + \frac{1}{12} \\[1ex]
\textit{Total} \quad & 90 + 9 + \frac{1}{2} + \frac{1}{4}
\end{array}
$$

Figure 1.7. Multiplying $(12 + 2/3)$ by $(7 + \frac{1}{2} + \frac{1}{4} + \frac{1}{8})$.

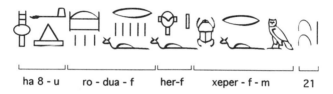

| ha 8 - u | ro - dua - f | her-f | xeper - f - m | 21 |

Figure 1.8. An Egyptian arithmetic problem. After René Taton, *Histoire du Calcul* (Paris: P.U.F., 1948).

when divided by 6, results in $3\frac{1}{2}$. The heap is therefore equal to $5 \times 3\frac{1}{2} = 17\frac{1}{2}$. That procedure is known as the *false position* method, and reveals the amazing arithmetical competence of the Egyptians, some four thousand years ago.

A TRIBUTE TO AHMES

The Rhind Papyrus is a copy attributed to a scribe named Ahmose, or Ahmes, of an earlier document, probably dating back to the Twelfth Dynasty of the Middle Kingdom (1900–1786 B.C.), the golden age of arts and crafts. The papyrus begins with these words:

Accurate reckoning. The entrance into the knowledge of all existing things and obscure secrets. This book was copied in the year 33, in the 4th month of the inundation season, under the King of upper and lower Egypt A-User-Ra, endowed with life, in likeness to writings made of old, in the time of the King of upper and lower Egypt Ne-Mat'et Ra. It is the scribe Ahmes who copies this writing.

It was not the custom for a scribe to identify himself. His personal glory stemmed from the practice of his art, and it behooved him not to appropriate his masters' science. It is conjectured that the author of the Rhind Papyrus was the great Imhotep, the pharaoh Zoser's architect and the builder of the Sakkara pyramid.[9]

> As for the learned scribes ... they did not make for themselves pyramids of copper with tombstones of iron. They were unable to leave an heir in the form of children who would pronounce their name, but they made for themselves an heir of the writing and instruction they hade made. (From a hieroglyphic inscription on an ostracon found in Deir el Medina).[10]

The Mesopotamians

> The Pythagorean theorem was well known
> to Babylonian mathematicians more than
> a thousand years before Pythagoras was born.
> (R. J. Gillings)[11]

Beginning with Georg Friedrich Grotefend in 1802,[12] Western scholars also turned their attention to the Mesopotamian cuneiform writings (from the Latin *cuneus*, meaning "wedge"). In 1854, twelve years after the publication of Champollion's *Dictionnaire Egyptien*, Assyriologist Hincks discovered examples of numeration on an

---

[9]It is not possible to close this brief history of Egyptian arithmetic without drawing the reader's attention to the exhaustive and systematic study conducted by Sylvia Couchoud of the University of Lyon, France; see Couchoud, *Mathématiques Egyptiennes* (Paris: Editions Le Léopard d'Or, 1993).

[10]Quoted in Andrea McDowell, "Daily Life in Ancient Egypt," *Scientific American*, December 1996.

[11]R. J. Gillings, *Introduction to Mathematics in the Time of the Pharaohs* (Cambridge: MIT Press, 1972).

[12]It is said that Grotefend succeeded in translating a cuneiform text following a challenge by his drinking companions.

astronomical tablet unearthed in Nineveh, and Sir Henry Creswick Rawlinson discovered other examples on a mathematical tablet found in Larsa. The decipherment of Mesopotamian numeration is largely owed to F. Thureau-Dangin and Otto Neugebauer, whose essential contributions lifted the veil not only on the principles of numeration, but on a wealth of other sophisticated mathematical achievements.

Around 2000 B.C., the Babylonians had mastered the art of solving quadratic as well as certain classes of cubic equations. They developed a place-value sexagesimal (base 60) number system, which eventually evolved into a true positional system with the introduction of a symbol for zero. They compiled star catalogs and records of planetary motion as early as 1800 B.C., thus laying the foundations of astronomy.

The Plimpton 322 Babylonian tablet (Figure 1.9), which dates back to 1800–1700 B.C., a millenium before Pythagoras, owes its name to its discoverer, and is kept at Columbia University. In 1945, it revealed to O. Neugebauer and A. J. Sachs that Mesopotamians were familiar with the so-called Pythagorean theorem. It is nowadays widely recognized that Pythagoras, who spent several years in Mesopotamia,

Figure 1.9. The Plimpton Tablet. Reproduced from Otto Neugebauer and Abraham J. Sachs, eds., *Mathematical Cuneiform Texts* (New Haven: American Oriental Society and American Schools of Oriental Research, 1945).

had learned from his Babylonian teachers the theorem that bears his name.

The most primitive additive number system was probably the scheme utilized by the Mesopotamian shepherd, who left the imprint of a pebble on a clay tablet for each sheep in the herd. Mesopotamians gradually developed more sophisticated representations, and by 3200 B.C., they had perfected several systems for measuring different kinds of objects—distances, areas, grain capacities, time, etc. One such system is shown in Figure 1.10.

| 1 | 10 | 60 | 600 | 3,600 | 36,000 |

Figure 1.10. Early Mesopotamian number icons.

These systems, elaborate as they were, remained additive for several centuries. Beginning around 2350 B.C., cuneiform symbols were gradually introduced, which were recorded by pressing a stylus against a soft clay tablet. Eventually, the repertoire of symbols was reduced to only two: the "nail," ⌷, and the "dovetail," ⌵, which denoted 1 and 10, respectively. These were grouped in tight little packets, which represented the numbers 1 to 59 (Figure 1.11a). A number larger than

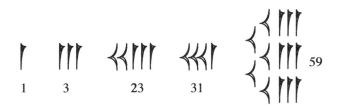

| 1 | 3 | 23 | 31 | 59 |

Figure 1.11a. Examples of cuneiform number packets.

59 consisted of a succession of packets, written from right to left. The contribution of any particular packet was equal to the packet's value multiplied by the power of 60 conferred by its position within the sequence (Figure 1.11b).

$$10\times60^3+ 1\times60^2+ 23 \times60 + 32 \times1 = 2\,165\,012$$

Figure 1.11b. The number 2,165,012.

Strictly speaking, the Mesopotamian system was not sexagesimal, as it is usually referred to. Rather, it constituted a periodic mixed radix system, of radices 10, 6, 10, 6, 10, 6, ...

The zero did not make its appearance until sometime between 300 B.C. and 700 A.D., in the form of a double slanted stroke. Meanwhile, scribes endeavoured to leave an empty space wherever a power of 60 was missing. Obviously, if the scribe was careless, the system could lead to serious ambiguities, as in Figure 1.11c. The spacing between

62	121	11	601

Figure 1.11c. Ambiguities in the Mesopotamian number system.

packets was therefore of considerable importance. Additionally, the power corresponding to the rightmost packet could be arbitrarily assigned by the reader, and the resulting ambiguity could be resolved only from the context. To evolve a truly positional system, a zero would have been required in lieu of a packet when the corresponding power of 60 was missing, as well as within the packet itself when either nails or dovetails were missing.

Mesopotamians also knew how to represent base 60 fractional numbers, in much the same way as we nowadays represent base 10 fractional numbers. However, the system suffered from a fundamental ambiguity stemming from the absence of zero. For example, (1, 38, 33, 36, 36) could represent

$$(1, 38, 33, 36, 36) = \frac{1}{60} + \frac{38}{60^2} + \frac{33}{60^3} + \frac{36}{60^4} + \frac{36}{60^5} = \frac{21\,288\,996}{60^5}.$$

That number is found on line 9 of the Plimpton Tablet, where it was intended to represent $21\,288\,996/60^4$.

With the introduction of a symbol for zero, the Mesopotamian system eventually evolved into a true positional, or place-value system, albeit with imperfections. Vestiges of the Mesopotamian system survive to this day in the measurement of time as well as in the division of the circle into degrees, minutes, and seconds.

The Mesopotamians performed their arithmetic operations by means of the abacus. They also resorted to elaborate multiplication tables (see accompanying illustration).

Cuneiform table of multiplication by 25.
Louvre Museum. Photo: Ch. Larrieu.

AN IMAGINARY BABYLONIAN EDUCATIONAL GAME

Let us imagine that history had taken a different, though plausible course, and that some ingenious Babylonian had invented little printing blocks embossed with the fifty-nine packet images, plus one blank (Figure 1.12a). The blocks would resemble little dominos, with

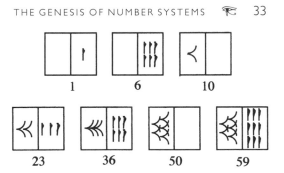

1   6   10

23   36   50   59

Figure 1.12a. Imaginary Mesopotamian imprinting blocks.

zero to nine nails embossed under the right-hand half, and zero to five dovetails under the left-hand side. Instead of resorting to the fastidious individual strokes, these blocks would be pressed against the soft clay. The number 129 898 832, say, would be imprinted as shown in Figure 1.12b.

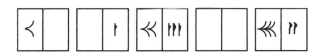

Figure 1.12b. Imprinting the number 129,898,832.

Upon noticing that the nails always occurred in positions of even rank, and the dovetails in positions of odd rank, our imaginative Babylonian could have justifiably concluded that two kinds of symbols were not really required, and retained the simple nail throughout. As a practical exercise, the reader is invited to open a box of dominos, discard the double six, and figure out which Babylonian numbers may be represented (see Figure 1.12c).

Figure 1.12c. An imaginary Babylonian domino game representation of the number 11,595,721.

Until the publication of *The Rhind Mathematical Papyrus* by A. B. Chace in 1927, and of *Mathematical Cuneiform Texts* by Neugebauer and Sachs in 1945, Western science had built an impressive mathematical edifice, and populated its pantheon with Greek geniuses. To the mind of almost every Western scientist and science historian, the genesis of mathematics dates back to Thales in the sixth century B.C., with its birthplace in the Ionian colonies of Miletus and Samos, in Asia Minor. Western classical culture thus traces its sources to the Hellenic civilization, with only a remote sense of kinship with Egypt and Babylon.

How many historians, over the centuries, considered the mathematics of Egyptians and Babylonians more akin to the scrawlings of children learning to write than to great literature! Fortunately that ill-educated view is not universally shared. Quoting James R. Newman,

> I am not impressed with the contention based partly on comparison with the achievements of other ancient people, partly on the wisdom of hindsight, that the Egyptian contribution was negligible, that Egyptian mathematics was consistently primitive and clumsy.... What is more to the point is to understand why the Egyptians produced their particular kind of mathematics, to what extent it offers a culture clue, how it can be related to their social and political institutions, to their religious beliefs, their economic practices, their habits of daily living. It is only in these terms that their mathematics can be judged fairly.[13]

The Greeks

As far back as the twelfth century B.C., the free spirit and boundless energy of the Greeks, combined with their exceptional seamanship, were at the root of the establishment of wealthy cities, where commerce and intellectual exchange began flourishing. The Greeks traveled extensively to Egypt and Mesopotamia, where they learned the sciences of the priests. Thales, the father of Greek science, spent a considerable number of years in these lands, and later encouraged Pythagoras to follow his example.

Mesopotamian and Egyptian sciences, despite their sophistication, were essentially oriented toward tangible applications. The Greeks, on the other hand, were engaged in the consistent pursuit of a coherent system, indeed, of beauty for its own sake. Greek scientists were interested only in abstract speculation, not in the world around them. They were interested only in *intelligible* objects, in other words objects created

---

[13]James R. Newman, *The World of Mathematics* (New York: Simon and Schuster, 1956).

by the intellect, as opposed to *sensible* reality, that which is perceived by the senses. Aristotle wrote, "*The science which has no tangible subject is above that which has.*" Of all the sciences, mathematics was regarded as the noblest activity, as attested by Plato's warning at the entrance of the Academy. However, Plato distinguished arithmetic, in other words the science of numbers, from logistics, the art of calculation, which he regarded as a lowly technique, not true science, and therefore devoid of the educational virtues that enhance one's intelligence.

Despite their extraordinary contributions to arithmetic, some of which are contained in this book, the Greeks were not able to produce a number system that posterity would eventually retain. They utilized a base 10 additive system, shown in Figure 1.13, where individual

α	β	γ	δ	ε	F	ζ	η	θ
1	2	3	4	5	6	7	8	9

ι	κ	λ	μ	ν	ξ	ο	π	ϟ
10	20	30	40	50	60	70	80	90

ρ	σ	τ	υ	φ	χ	ψ	ω	ϡ
100	200	300	400	500	600	700	800	900

,α	,β	,γ	,δ	,ε	,F	,ζ	,η	,θ
1000	2000	3000	4000	5000	6000	7000	8000	9000

Figure 1.13. Greek numbers.

letters of the alphabet stood for the nine digits 1 to 9, as well as for the numbers 10, 20, 30, ..., 100, 200, 300, ..., 1000, 2000, 3000, ... The system was difficult to use, even when performing elementary arithmetic operations such as addition and multiplication. Thanks to the sand abacus, the Greeks could nonetheless perform very complex calculations. They also resorted to multiplication and division tables, which were so difficult to learn that only experienced mathematicians could master their use.

Upon adopting the Mesopotamian system for their astronomical calculations, its relative simplicity might have reasonably led the

Greeks to reflect upon the virtues of the positional system. But they failed to appreciate its merits and went on using the letters of the alphabet, a practice that was probably also responsible for their failure, until the twilight of their great civilization, to invent algebra. Gauss, one of the greatest mathematicians of all time, had an immoderate admiration for Archimedes, of whom he said, "How could he have missed the discovery of our present positional system of writing numerals? To what heights science would have risen by now, if only he had made that discovery!"

Because it is typical of the rhetorical statements of the time, let us consider for a moment, without necessarily attempting to understand it at this point, the famous statement by Eudoxus which founds his definition of irrational numbers:

> Magnitudes are said to be in the same ratio, the first to the second and the third to the fourth, when, if any equimultiples whatever be taken of the first and the third, and any equimultiples whatever of the second and the fourth, the former equimultiples alike exceed, are alike equal to, or are alike less than, the latter equimultiples taken in corresponding order.[14]

That statement is difficult to understand for modern-day students of mathematics, who are accustomed to algebraic symbols such as a, b, m, n, etc., and relations between them. It was certainly not with a view to deliberately making their mathematical statements inaccessible that ancient Greeks resorted to such convoluted language. They simply had no other language at their disposal, and it was not until Diophantus (ca. 250) that they began using rudimentary algebra. According to Tobias Dantzig,

> Greek algebra before Diophantus was essentially rhetorical. Various explanations were offered as to why the Greeks were so inept in creating symbolism. One of the most current theories is that the letters of the Greek alphabet stood for numerals and the use of the same letters to designate general quantities would have obviously caused confusion.[15]

---

[14]See T. L. Heath, *The Thirteen Books of Euclid's Elements* (Cambridge, 1908).

[15]Tobias Dantzig, *Number, the Language of Science*, 4th ed. (Garden City, N.Y.: Doubleday, 1954), p. 79.

The introduction of algebra by the Arabs allowed mathematicians to replace verbal statements by symbols assembled according to strict construction rules. What was referred to by al-Khawarizmy as *al-Moqabala*, literally meant the *opposition* of symbols on either side of an equality sign. In all probability, the unfortunate choice of the alphabet by the Greeks, in addition to their lack of interest in earthly matters, not only prevented them from discovering the positional number notation, but also from inventing algebra. How many such accidents must have occurred in the course of history! How many times did early mathematicians unwittingly come upon a bifurcation, and by some contingent happenstance, reminiscent of Darwinian evolution, chose one branch of the alternative instead of the other, letting history take its irreversible course? The zero was one such bifurcation, and algebraic symbolism another.

The decline of Greek civilization began following Archimedes' death in 212 B.C. It is often maliciously observed that the only Roman contribution to science was the slaying of Archimedes by a Roman soldier upon the fall of Syracuse. That view is not entirely fair, and the decline of Greek science should not be attributed solely to the Roman conquest. Its seeds must be sought within the Greek civilization itself. According to L. Brunschwig, "The science of antiquity lacked what we regard today as the very condition of knowledge: the connection between calculation and physical experimentation."[16] Greek science contained the germ of its own destruction. Wrote P. Boutroux, "The exclusive character of its field of action, the aesthetic character of its preoccupations were bound to eventually stop its development."[17]

The Romans, on the other hand, were not interested in intellectual speculation. In their mathematical treatises, they did away with proof, the very substance of mathematics, and provided only practical formulae. Cicero observed that "to the Greeks nothing was more illustrious than mathematics, but we have limited the realm of this art to measurement and practical reasoning." Roman numeration, which to this day is used to engrave dates upon the pediments of Catholic churches, sometimes to number a book's chapters, and in a few other special contexts, did not authorize any measure of written calculation. The Romans resorted exclusively to the abacus.

---

[16] L. Brunschwig, *Les Etapes de la philosophie mathématique* (Paris, 1947).
[17] P. Boutroux, *Les Principes de l'analyse mathématique* (Paris, 1914–1919).

It is impossible to leave that region of the world without reflecting upon the immense loss suffered by humankind upon the burning of the Alexandrian Library, which was built by Ptolemy I, shortly after the conquest of Egypt by Alexander the Great and the founding of Alexandria. It contained no less than thirty thousand volumes, reflecting millenia of accumulated knowledge. Among these were the thirty volumes of Egyptian history written during the reign of Ptolemy I by Menathon, an Egyptian priest. Menathon, upon writing his books, no doubt had access to innumerable documents kept in Egyptian libraries and temple archives.

The library burned in the year 47 B.C., during the conquest of Egypt by Julius Caesar, probably by accident. As was customary in those days, copies of major works were kept in different places, and a copy of Manethon's opus survived for four hundred more years in the temple of Serapis, also situated in Alexandria. That temple was burned to the ground in 391 by the troops of the Byzantine emperor Theodosius I, who ordered the closure or destruction of all pagan temples. Egyptian priests thus gradually disappeared, and with them the history, culture, and alphabet of Ancient Egypt. Nonetheless, the Byzantine scribes dedicated themselves to copying ancient texts, albeit not always faithfully. The fall of Constantinople in 1453 marks the end of the Byzantine era. Fortunately many scholars were able to flee to Italy, rescuing several valuable manuscripts from oblivion. These would provide the foundations of the rebirth of Western science. As these lines are being written, the Egyptians, with the help of international cultural organizations, are breaking ground for the new Alexandrian library, which is due to open its doors at the dawn of the third millenium.

The Mayas

The name Maya refers to the Central American tribes who inhabited the regions known today as Guatemala and Yucatan. Their civilization flourished from the third to the eighth century, and again from the tenth to the twelfth century. Recently discovered artifacts indicate that their civilization probably dates back to the fifth or sixth century B.C. It would therefore span the period stretching from Pericles in Athens to the Middle Ages in Europe. The ruthless Spanish invasion of 1524 fell upon a civilization that had already begun to decline, as all ancient civilizations eventually did. That decline was

caused by the failure of the Mayan political system, by the deliques-
cence of their elites, and by famine. The constructions, carvings,
and artifacts left by the Mayas reveal a great degree of advancement,
notably in the field of astronomy, which was central to their culture
and beliefs. The Mayan year, inherited from previous civilizations,
Zapotec and Olmec, comprised eighteen months of twenty days each,
as well as five additional days, reminiscent of the ancient Egyptian
epagomenal days.

Traces of Mayan calculation may be notably found in the four
copies of the Mayan Codex (Plate 7) that survived the Spanish invasion, which are preserved in Mexico, Madrid, Paris, and Dresden. The
calculations contained in the codex all pertain to the Mayan calendar,
which accounts for a strange singularity in their number system, the
intrusion of the radix 18 within an otherwise vigesimal base.

The Mayan number system is a *place-value*, or *positional*, sys-
tem. Its base, written from right to left, is $(\ldots m_3, m_2, m_1, m_0.) = (\ldots 20, 20, 18, 20.)$. If we allow the symbols δ_0, δ_1, δ_2, $\delta_3, \ldots$ to
denote a number's successive digits, where δ_0 is the lowest-rank digit
(representing the units), we may assign each digit δ_i a *weight* π_i, as in
Table 1.1. That weight determines the digit's contribution to the number. The zero, an essential ingredient of the system, was represented
by a variety of glyphs resembling a seashell (Figure 1.14a).

TABLE 1.1
Digits δ_i, base radices m_i, and weights π_i
in the Mayan system.

δ_i	m_i	π_i
δ_4	20	$144000 = 20 \times 18 \times 20 \times 20$
δ_3	20	$7200 = 20 \times 18 \times 20$
δ_2	20	$360 = 20 \times 18$
δ_1	18	$20 = 20$
δ_0	20	1

$$N = \delta_0 + 20\delta_1 + (20 \times 18)\delta_2 + (20 \times 18 \times 20)\delta_3 + \cdots$$

$$= \delta_0 + 20\delta_1 + 360\delta_2 + 7200\delta_3 + \cdots$$

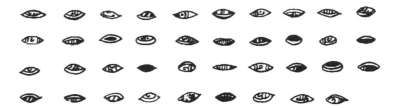

Figure 1.14a. Various Mayan glyphs representing zero. Reproduced from *Histoire universelle des chiffres*, by George Ifrah (Paris: Robert Laffont, 1994).

Our present-day decimal system comprises a collection of no more than ten ciphers, one for each number from 0 to 9. Notwithstanding the 18 oddity in the second rank, the Mayan system might have contained twenty different glyphs, one for each number from 0 to 19.

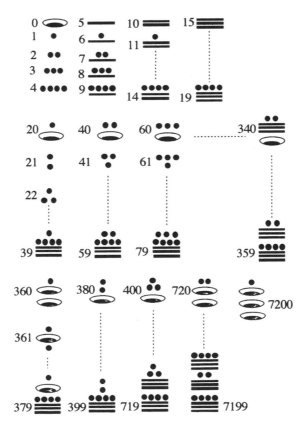

Figure 1.14b. Mayan numbers.

Instead, the Mayans resorted to a futher breakdown of each individual vigesimal digit into a combination of up to four dots representing one unit each, and up to three bars representing five units each, as in Figure 1.14b. The bar and dot alignments were indifferently horizontal or vertical, as in the fragment of the Mayan Codex reproduced in Plate 1.7.

In our decimal system, by virtue of its uniformity (all base radices are equal to 10), multiplying a number by 10 simply entails the addition of a zero on the number's right. Multiplying a Mayan number by 20 was not an easy matter, and did not necessarily boil down to adding a zero on the number's bottom. For example, whereas $721 \equiv (2, 0, 1.)$ and $20 \times 721 = 14420 \equiv (2, 0, 1, 0.)$, we find that $192 \equiv (9, 12.)$ and $192 \times 20 = 3840 \equiv (10, 12, 0.)$. The numbers between parentheses are written here from right to left. Convenient as the system may have been for astronomers, the 18 singularity probably constituted an insurmountable intellectual hurdle for the average person.

TWO CURRENT NUMBER SYSTEMS

> It is India that gave us the ingenious method of expressing all numbers by means of ten symbols, each symbol receiving a value of position as well as an absolute value; a profound and important idea which appears so simple to us now that we ignore its true merit. But its very simplicity and the great ease with which it has lent to all our computations put our arithmetic in the first rank of useful inventions; and we shall appreciate the grandeur of this achievement the more when we remember that it escaped the genius of Archimedes and Apollonius, two of the greatest men produced by antiquity.
> (Simon Marquis de Laplace)

It is quite extraordinary that neither the Egyptians nor the Greeks were able to translate the empty column of the abacus into a symbol representing zero, or nothingness. According to Karl Menninger, the conceptual difficulty may have been that the "zero is something that must be there in order to say that nothing is there."[18] To the ancient

---

[18] Karl Menninger, *Number Words and Number Symbols* (Cambridge: MIT Press, 1970).

Greeks, numbers did not only have an existence of their own; they were everything, and everything was number. "All things which can be known have number," said Philolaus, "for it is not possible that without number anything could be conceived or known." "Nothing" was not a number and could not be endowed with individual existence. It could not have a name, and no symbol could represent it. It took several centuries, and the intercession of the Arabs, for the Western world to finally pay attention to the Hindu numeration system, complete with its zero.

The Hindus

Hindu mathematicians were probably better prepared culturally and philosophically to cope with the profound paradox of naming the unnamable, and representing the absence of substance by substance. Indians were uniquely gifted for abstraction, and number provided them with the tools of a profoundly virtuous exercise. Honoring divinity was often done by endlessly counting, the noblest of human rituals, and it was not uncommon to find poetry interspersed within Sanskrit scientific tests.

Vasumita, a Buddhist monk of the first century, defended the idea that if a substance, upon traversing the three "states" (Past, Present, Future), is declared "other" every time it enters a new state, the change it undergoes must be regarded as deriving from the "otherness" of the state, not the substance. That may have provided the philosophical underpinning for a system where a scribbling on the dust would come to be assigned different values depending not only on the scribbling itself, but on where it fell on the ground.

India is often thought of as a faraway culture which evolved from within, without being subjected to external influences. Nothing could be farther from the truth. Following the conquest of Darius and the Alexandrian expedition, India was exposed on the one hand to the realistic, practical influences of the East, and on the other to the more intellectual influences of the Greeks. Their intellectual pursuits relied on intuition as often as they relied on logical deduction.

Using a symbolic transcription of the abacus, the Hindus acquired an extraordinary proficiency in numerical computation. They not only invented the numeration system that we presently use, but also gave birth to algebra and trigonometry. Most notable among them were Aryabhata (sixth century), Brahmagupta (seventh century), and Bhaskara (twelfth century), who discovered an elegant proof of the theorem of Pythagoras, to which we shall return.

The Arabs

> Now, however, it is necessary that we should
> demonstrate geometrically the truth of the same
> problems which we have explained in number.
> (Mohammed ibn Mussa al-Khawarizmy,
> *Al-Jabr W'Al-Moqabala*)

What we nowadays refer to as ancient Greece was in reality a mixture of nationalities, not always harmonious, brought together by conquest, and by a common culture. Thales and Pythagoras were born on the eastern Aegean coast, and the latter established his quarters in Italy. Euclid spent most of his productive life in Alexandria, as did numerous other great mathematicians, including Hero, Ptolemy, and Diophantos. Theodorus was from Cyrena, Eudoxus from Cnidus, Plato established the Academy in Athens, etc. Hence the metaphor by André Sigfried: "the Mediterrannean is the world's only liquid continent." Similarly, the people referred to as the Arabs came from an extensive geographical area, brought together by Islam and a common language, Arabic.

When the Prophet Mohammed was born in Mecca, around 570, the Arabian peninsula was mostly inhabited by illiterate nomad tribes. Mohammed preached for several years at Mecca, advocating a monotheistic religion. Islam, literally meaning "submission" (to the will of God) is based on the Koran, the revealed word of God, as well as the Hadith, the Sayings, and the Sunna, the manner of life of Mohammed. Following an assassination plot, Mohammed moved to Medina in 622, in an episode known as the Hegira, marking the beginning of a new era of military, spiritual, and cultural expansion. Shortly thereafter, he founded an Islamic state, with Mecca as its capital. Following Mohammed's death in 632, his successors established an immense empire, stretching from Spain to the confines of China.

The Arabs absorbed the intellectual achievements of the conquered territories with extraordinary eagerness, and ushered in an era of enlightenment, with Baghdad as its focal point, which eventually led to the European Renaissance. India was among the most mathematically advanced regions of the empire, and provided the Arabs with the foundations of what was to be later known as Arabian mathematics. Around 766, the *Siddhanta*, a book of astronomy and mathematics was brought to Baghdad, and later translated into the Arabic *Sind-Hind*. Around the same time, Ptolemy's *Tetrabiblos*, an astrological textbook, was translated from Greek into Arabic. During the flourishing Abbassid period, the Caliph al-Ma'mun had a vision of Aristotle in a dream, and thereupon ordered the translation of every major Greek book available. Thus were translated Ptolemy's *Almagest* and Euclid's *Elements*. The Europeans later discovered Euclid's work through its Arabic translations (Plate 8).

It is al-Ma'mun who established Beit el Hekma, the House of Wisdom, in Baghdad, where erudites flocked in from all over the empire. One of the learned scholars of that noble institution was the Persian Mohammed ibn Mussa al-Khawarizmy (d. 850), named after Khwarazm, a city of Uzbekistan. Al-Khawarizmy made several fundamental contributions to astronomy and mathematics, and is remembered for his momentous *Al-Mukhtassar fi hisab al-jabr w'al-moqabalah*. The Arabic *al-jabr*, from which the word "algebra" derives, means "restoration," or "reduction," as one reduces a fracture. (To this day the local bone-setter in some Spanish provinces is known as the algebrista.) *Al-moqabalah* means "opposition," probably referring to the two sides of an equation. The word comes from the verb "*qaabala*," which means "to face," or "to correspond," whereas the modern Arabic word for equation, *mo'adalah*, comes from the verb "'*aadala*," meaning

فأما الأموال والجذور التى تعدل العدد فثل قولك

مال وعشرة أجذاره يعدل تسعة وثلاثين درهما ومعناهأى مال اذا زدت عليه مثل

عشرةأجذاره بلغ ذلك كله تسعة وثلاثين . فبابه^(٢) أن تنصف الأجذار وهى فى

هذه المسئلة خمسة فتضربها فى مثلها فتكون خمسة وعشرين فتزيدها على التسعة

والثلاثين فتكون أربعة وستين فتأخذ جذرها وهو ثمانية فتنقص منه نصف

الأجذار هو خمسة فيبقى ثلاثة وهوجذر المال الذى تريد والمال تسعة .

Figure 1.15. Passage from *Al-Jabr w'al-Moqabalah*, excerpted from the Arabic version by A. M. Ashrafa and M. M. Ahmed (Cairo, Egypt, 1968). Reproduced from J. L. Chabert, ed., *Histoires d'algorithmes* (Paris: Editions Belin, 1994).

"to compensate, arrange, balance, equal." The word "algorithm," which results from a contamination of the name al-Khawarizmy by the Greek word *arithmos*, meaning "number," has come to represent any iterative, step-by-step procedure. Figure 1.15 is an excerpt from al-Khwarizmy's book outlining the algorithm for the solution of the equation $x + 10\sqrt{x} = 39$. The following is a transcription of the algorithm contained in the text:

$$\text{to solve} \quad x + 10\sqrt{x} = 39$$
$$\text{do} \qquad \frac{10}{2} = 5$$
$$5^2 = 25$$
$$25 + 39 = 64$$
$$\sqrt{64} = 8$$
$$8 - 5 = 3$$
$$x = 3^2 = 9.$$

The reader may verify, putting $y = \sqrt{x}$, that the problem consists of solving $y^2 + 10y - 39 = 0$. The equation's real positive root is $y = \dfrac{-10 + \sqrt{100 + 156}}{2} = 3$; hence $x = 9$.

The Decimal Number System

Novem figure Indorum he sunt 9 8 7 6 5 4 3 2 1.
Cum his itaque novem figuris, et cum hoc signo 0,
quod Arabic cephirum appellatur, scribitur quilibet numerus.
(Leonardo of Pisa, *Liber abaci*)

The greatest contributor to the introduction of the Hindu-Arabic system to the West was without any doubt Leonardo of Pisa (1180–1250), otherwise known as Fibonacci. The son of Bonaccio, Pisan governor of Bougie in Algeria, Leonardo took up the study of computation, *studium abaci*, with an Arab whom he described as a magnificent teacher. Fibonacci published his momentous *Liber abaci* in 1202, beginning with the words: "The nine numerals of the Indians are 987654321. With these nine figures and with that sign 0, which in Arabic is called cephirum, any desired number can be written." *Liber abaci*, literally meaning "the book of the abacus," might be construed as a misnomer, as the book is entirely devoted to the Arabic number system and its use. At that time, however, *ars abaci*, the art of the abacus, had come to apply to the general art of computation, regardless of which system was being used.

The newly discovered virtues of the zero were so overwhelming that its Arabic name, *sifr*, which was transformed by Jordanius Nimerarius into *cifra*, came to represent the whole set of ten digits, or "ciphers" in English, and *chiffres* in French, thus raising ambiguities at times. As to the word "zero," it came from the graceful Latinization of *sifr* into *zephyrium*, reflecting the ethereal quality of that symbol of absence (Figure 1.16). (In contemporary Egypt, an individual despised by his or her peers, is metaphorically referred to as *sifr 'ala al-yassaar*, literally meaning "zero on the left-hand side," or as we would say today, "nonsignificant zero.")

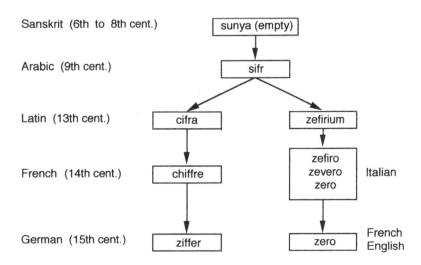

Figure 1.16. The zero. After Karl Menninger, *Zahlwort und Ziffer*
© Vandenhoeck & Ruprecht in Göttingen, 1958.

Following the translation of al-Khawarizmy's book into *De numero Indorum* (Concerning the Hindu art of reckoning), the ciphers were referred to as *arabic numerals*, whereas the author himself willingly admitted his Indian source, probably Brahmagupta. Figure 1.17a shows

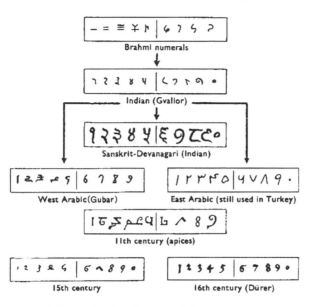

Figure 1.17a. The family tree of the Indian numerals.
Reproduced from Karl Menninger, *Zahlwort und Ziffer*
© Vandenhoeck & Ruprecht in Göttingen, 1958.

how the Sanskrit numerals evolved into the fifteenth-century symbols. Figure 1.17b illustrates the manner in which Albrecht Dürer stylized the Arabic numerals in the sixteenth century. Figure 1.17c shows a page from a modern Egyptian elementary arithmetic textbook.

The new system, though extremely appealing to astronomers, as well as to merchants and bankers, whose sons flocked to Italy from every corner of Europe to learn the new art of computation, met with

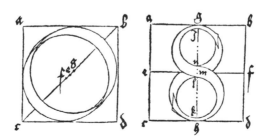

Figure 1.17b. Stylized numerals by Albrecht Dürer.
Reproduced from the French translation of Dürer's *Geometry*
by Jeanne Peiffer (Paris: Editions du Seuil, 1995).

أُكْتُبْ حَاصِلَ الطَّرْحِ الصَّحِيح   اكْتُبْ حَاصِلَ الْجَمْعِ الصَّحِيح

$$٦ - ٧ \qquad ٤ + ٣$$

$$٥ - ١٠ \qquad ٢ + ٦$$

$$١ - ٤ \qquad ٥ + ١$$

$$٣ - ٩ \qquad ٣ + ٢$$

$$٧ - ٨ \qquad ٩ + ١$$

Figure 1.17c. Facsimile of a page from a modern Egyptian elementary arithmetic textbook.

considerable opposition from the establishment of the time. In 1299, the year following the publication of Marco Polo's *Book of Travels*, the City Council of Florence issued an ordinance forbidding the use of the new numerals in official accounting books. These numerals, the council members argued, could not be connected together without lifting the pen from the paper. They ruled that the old figures could not be as easily falsified as the new ones, which could be turned into different figures without difficulty. Deep into the fifteenth century, resistance was still very strong, as attested by the order handed down by the Frankfurt mayor, that the master calculators abstain from using the new numerals. It took centuries for the new system to eventually prevail, as the dark Middle Ages were coming to a close, and enlightenment was rising on the horizon.

When Gutenberg perfected the printing press (around 1454, the year of the Bible printing), the Hindu-Arabic system was afforded such unprecedented visibility that while the rest of Europe referred to ciphers as Arabic numerals, the Germans eventually deemed the digits worthy of being called *Deutsche*! Be that as it may, the *Algoristic* textbooks, advocating the new system almost three hundred years after Fibonacci's *Liber abaci*, were among those most in demand, and were printed in large numbers. The first book of arithmetic ever to be printed was the *Treviso*, a list of rules for performing common calculations, published in 1472. The first book of immense mathematical significance was printed by Johannes Campanus in 1482. It was the Latin translation of Euclid's *Elements* from its earlier Arabic translation.

Figure 1.18. Gregor Reisch's *Margarita Philosophica*.

The mood of that period is beautifully captured by an illustration
that appeared in Gregor Reisch's *Margarita philosophica* (Philosophical
pearl), published in 1503 (Figure 1.18). A female figure personi-
fies arithmetic—*Typus arithmeticae*—with the progressions 2-4-8 and
3-9-27 adorning her garment. To her right, the better of the two flanks
in more than one culture, is seated an obviously content Boëthius, who
seems to have successfully performed some operation using digits.
To her left, a dismayed Pythagoras mulls over his line board abacus,
which displays the numbers 1,241 and 82. In that allegory of the feud
which opposed abacists and algorists, the latter were championed erroneously, but perhaps poetically, by Manlius Boëthius, who lived in the
sixth century, and is credited with the classification of numbers into
nine *digiti*, or "fingers."

But the feud lingered on into our 1950s, when a contest was staged between a Japanese calculator, aided only by his soroban, the Asian abacus, and an electronic digital calculator of the day, which "calculated with the speed of light." Against all odds, the Japanese contestant exceeded the "speed of light"! In fairness to the computer, the race was really one between the ergonomic merits of two rival input-output devices, and in those days, entering and exiting a computer was not achieved as convivially as it is done nowadays. The dexterity of the computer operator was no match for that of the Japanese abacist, and no sooner had the latter entered his numbers than the result was available.

Fractional Numbers

Teaching how all computations that are met in
business may be performed by integers alone without
the aid of fractions... (Simon Stevin, *Die thende*)

Anyone can understand that the fraction $1/3$ represents the third part of something, as in our daily life, we experience the division of a whole into so many parts. Understanding, on the other hand, that $1/3$ is equal to the sum of an infinite number of ever-shrinking fractions— $3/10$, $3/100$, $3/1000$, ...—is a much more difficult task, and ancient Greeks of the intelligence of Zeno the Eleate were not able to accept that $1/2 + 1/4 + 1/8 + 1/16 + \cdots$ could add up to 1. It took several centuries for the positional representation of fractional numbers to prevail, and early calculators resorted to elementary fractions, such as $1/3$ or $2/5$, etc. The positional representation of integers was itself quite difficult to comprehend, let alone that of fractional numbers. The latter entails conceptual difficulties, which are embedded in the essentially infinite nature of the positional representation.

With recent translations of Mesopotamian cuneiform tablets, it is now clear that Babylonians used the successive powers of $1/60$ to weigh sequences of integers representing fractional numbers. The system was somewhat ambiguous, however, as the lowest power of $1/60$ was not always clearly indicated, and had to be decided upon from the context. Be that as it may, their system was the best available for centuries, and Leonardo of Pisa himself, the staunch defender of the base 10 system who introduced Hindu-Arabic numeration to the West in the thirteenth century, had to fall back on the Mesopotamian sexagesimal system to accurately compute the root of a cubic equation, in the course of a mathematical tournament.

Three centuries after Leonardo, the use of the positional system for the representation of fractional numbers was explained by Christoff Rudoff, an enlightened Viennese Renaissance man, in *The Rules of Coss* and the *Reckoning Manual*, published in Augsburg in 1530. Half a century later, in 1585, Simon Stevin published *Die Thiende* in Dutch, shortly followed by *La Disme*, its French translation, in which he gave a systematic presentation of the rules for handling decimal fractions (Figure 1.19). Simon Stevin (born in Bruges in 1548, died in 1620) was a Dutch mathematician of unusual genius who, like his contem-

Figure 1.19. Page 13 from Stevin's *Die Thiende*

porary Galileo, fought the paradigms of his time. In 1586, he showed that two different weights dropped together from the same height hit the ground at the same time, and that the pressure exerted by a liquid on an object depended upon the liquid's height. He notably postulated the impossibility of perpetual motion, which inspired Huygens's first statement of the principle of conservation of energy.

Before Simon Stevin's time, with a few notable exceptions, fractional numbers were represented by decimal fractions, as in $12^{345}/_{1000}$. Stevin advocated using the form $12 \ 3^{I} \ 4^{II} \ 5^{III}$, and, following innumerable variations on that theme, Balam introduced in 1653 the notation 12:345, the precursor of our present-day notation, which, in addition to its great simplicity, makes it possible to "float" the decimal point back and forth, multiplying or dividing a number by any power of 10 at will.

Modern-day computer experts, who are also expert at coining misnomers, have elected to freeze a number's decimal point in a fixed location and accompany its representation with its exponent, namely, the power of 10 by which it must be mutiplied. They refer to that construction as floating point arithmetic, when it should be more appropriately called frozen point arithmetic.

Uttering Versus Writing

Or else that numbers wax till ten they reach,
And then from one begin their rythm anew.
(Ovid, *Fasti* 3)

As children learn to count, they encounter no major difficulty on the road from one to nine, as they skip from one finger to the next. But suddenly, a major intellectual chasm opens up. One step beyond ten, and they are thrust onto another plateau, upon which they must again count from one to nine, while making reference to that new plateau with every finger they skip anew. Whereas an Arabic-speaking child would simply utter the equivalent of "one'ten, two'ten, three'ten...," the less fortunate French-speaking child learns to recite such monstrous number words as *onze, douze, treize, quatorze, quinze, seize...,* whereupon the child breathes a sigh of relief and falls back on the friendlier *dix-sept, dix-huit, dix-neuf....* The next plateau is reached with the count of twenty, etc. Despite the numerous linguistic hindrances strewn by insensitive educators along the road, such as *quatre-vingt-douze*, literally meaning "four score twelve," the child eventually reaches ninety-nine, where a new logical hurdle awaits. The historical date of 1789 is read "seventeen hundred, four score, nine"! The Anglo-Saxons do not fare much better with their "seventeen, eighteen, nineteen," as attested by the South African rhyme "...each, peach, muskydom/Tillatah, twenty-one," which is most probably a deformation of "Each speech must be dumb/Till I tell twenty-one." Upon learning to recite the Declaration of Independence, every American child begins with the historic phrase "Four score and seven years ago."

Despite the universal acceptance of the Hindu-Arabic positional system, the additive (truly, *addictive*) system lingers on to this day when it comes to *uttering* number names. The positional system uses no more than nine symbols, each of which has a name and a specific symbol, whereas decimal additive systems have names as well as specific symbols for units and powers of 10, commensurate with the largest magnitudes that our ancestors were susceptible of encountering in their daily lives.

Today, the specific Egyptian and Roman symbols for powers of 10 have altogether disappeared, but their names survive. The number 19,008 is not read "one nine zero zero eight," but "nineteen thousand eight." Had not the number word "Myriad" disappeared, we would probably be saying "one myriad, nine thousand eight." The Japanese language still uses *man* for "ten thousand,"[19] and "one million" is uttered *Hyaku-man*, literally meaning "100 man."

In the Western world, those who are aware of the Arabic origin of our present-day number system often express surprise at the fact that, though Arabic is written from right to left while most other languages are written in the opposite direction, the Arabs write their numbers "the way we do." The truth is that it is we who write our numbers the way they do. The difference is that we utter them from left to right, beginning with the highest power of ten, while in many, if not all Arab countries, numbers are uttered from right to left, beginning with the units. For instance, 1999 would be read "nine and ninety, nine hundred and one thousand."

No wonder, then, that our chidren are bewildered by such inconsistencies. Only the gifted can read and take number dictation by

---

[19]Derived from the Chinese *wan*.

the time they are fluent in the alphabet and elementary word construction, not to mention the equally innumerable inconsistencies in orthography and grammar.

Vestiges of ancient additive systems will probably remain in use forever, because of the relative conciseness of names given to groupings, in addition to the intuitive sense of quantity that is conveyed by such names, compared to ordered sequences of unit number names. Very large, nameless numbers, however, are verbally conveyed by falling back, at least partially, on the positional system. One cubic centimeter of water is declared to contain "ten to the power twenty-three" molecules, or, if one really wants to impress one's audience, "one followed by twenty-three zeros," obviously a hybrid of the additive and positional systems.

Units

> In view of the great usefulness of the decimal division, it
> would be a praiseworthy thing if the people would urge
> having this put into effect, so that . . . the State would declare
> the decimal division of the large units legitimate to the
> end that he who wished might use them. It would
> further this cause also, if all new money should be
> based on this system of primes, seconds, thirds, etc.
> (tenths, hundredths, thousandths).
> (Simon Stevin, *Die Thiende*)

It took exactly two hundred years, and a revolution, for the French to heed Simon Stevin's wise advice, and two hundred more years for the British to follow. For several centuries following the universal acceptance of the positional decimal system, several countries, not the least of which was Great Britain, were still using awkward unit systems for measuring objects, often depending on the nature of the object itself. Distance was measured in miles, feet, inches, and binary fractions thereof, and currency units were the penny, shilling, pound, etc. Along with the "three R's," English children had to spend considerable time and effort converting within each unit system, and between systems. (Given the price in pence of a length of string given in feet, the child was required to calculate in pounds, shillings, and pence the price of a length given in miles, feet, etc.) Meanwhile, contemporary French pupils, enlightened by the French Revolution, were relieved from these complicated chores, and learned that *deci* meant one-tenth, *centi* meant one-hundredth, and so on, whether they were dealing with weights, distances, areas, volumes, or other units. Plate 9 is a reproduction of a revolutionary print boasting of the new units—the liter, the gram, the meter, the are (100 square meters), the franc, and the stere (a measure of solid volumes). And yet, even in France, guardian of the platinum one-meter standard in Sevres, the present-day measurement of time is duodecimal when counting hours, sexagesimal when counting minutes and seconds, and decimal when counting fractions of a second!

The Binary Number System

A method for signifying one's mind to any distance by
objects that are either visible or audible, but capable of
two differences, such as fireworks, bells, cannon....
(Francis Bacon)

Sir Francis Bacon (1561–1626), a herald of the New Philosophy that purported to accompany the emergence from the dark Middle Ages into an era of enlightenment and modern thought, "imagined a method for signifying one's mind to any distance by objects that are either visible or audible, but capable of two differences, such as fireworks, bells, cannon." His system consisted of reducing the alphabet to two letters only, say, *a* and *b*. Every letter in the extended alphabet was represented by a specific configuration of the two letters (with the exception of *j* and *u*). His biliteral alphabet was so constructed that by replacing *a* with zero and *b* with one, the twenty-four remaining letters exactly corresponded, in alphabetical order, to the numbers 0 to 23 in our present-day binary notation.

The binary system, whose widespread use began with early accounting machines, was suited to a universe of punched cards and electromechanical relays. As in the original Jacquard loom, a prescribed position in a piece of cardboard is or is not punched. Similarly, a relay is either open or closed. Furthermore, the underlying *logic* governing the design of decisional circuits that route and process information is itself amenable to binary treatment and implementation. In 1932, the Frenchman Raymond Valtat patented, in Germany, a calculating apparatus predicated on the binary system, and in 1936 published an article entitled "Machine à calculer fondée sur l'emploi de la numération binaire." That same year, Alan Turing, the short-lived British mathematical genius, designed an electromechanical binary multiplier, and in 1938, the American George Stibitz built a binary adder using electromechanical relays. In 1946, A. W. Burks, H. H. Goldstine, and John von Neumann published a memorandum in which they advocated abandoning the decimal representation in favor of the binary system, and in 1959, W. Buchholz wrote an internal IBM research paper entitled "Fingers or Fists," a clever metaphor opposing the two systems, proposing a comparative analysis of their respective merits. In the universally accepted von Neumann model, numerical data as well as information regarding the handling of these data are expressed and treated in like fashion within the machine.

Computers have thus evolved into enormously intricate webs of electrical paths, where, at any given instant of time, current flows or does not flow. By virtue of the recursive character of the positional number system, and of the algorithmic procedures that are utilized throughout, the computer's structure itself is largely recursive: identical sub-assemblies of the machine are repeated over and over, as abundantly as technology allows at a tolerable cost.

It is indeed remarkable that the invention of the transistor in the late 1940s, and the unprecedented complexification that was subsequently enabled by large-scale integration, preserved and amplified the initial reliance on devices that are either open or closed. To be sure, many attempts were made in the laboratory to develop multistable devices, capable of assuming more than two equilibrium states, and an equally large number of interesting attempts were made by several contributors (including this author!) to further the mathematics of many-valued logic.[20] As the world's first electronic calculator was being designed by Eckert and Mauchley in 1945–1946 at the Moore School of Electrical Engineering, the "balanced ternary system" was regarded as a possible candidate, along with the binary system. That system, which uses digits −1, 0, 1, had been invented one hundred years earlier by Léon Lalanne, a French designer of arithmetical engines, and published in the *Comptes rendus* (Paris, 1840). Wrote

---

[20]M. J. Gazalé, *Les structures de commutation à m valeurs et les calculatrices numériques* (Paris: Gauthier Villars, 1959); D. C. Rine, *Computer Science and Multiple-Valued Logic* (Amsterdam: Elsevier, 1984).

Donald E. Knuth, in his famous 1969 *The Art of Computer Programming*: "*So far no substantial application of balanced ternary systems has been made, but perhaps its symmetric properties and simple arithmetic will prove to be quite important some day (when the 'flip-flop' is replaced by a 'flip-flap-flop')*."[21] Laudable as these efforts may be, the binary system has so far victoriously resisted their assaults. It is highly reliable, easy to conceive, and economical to implement. But one never knows. Maybe the trailblazing development of Fuzzy Logic[22] will breathe new life into multivalued logic. Be that as it may, the robustness and versatility of the so-called digital technologies has allowed them to permeate every aspect of modern information systems. The art of computers and that of telecommunications have gradually merged into a single field of human activity, later joined by broadcasting, electronic games, and a wealth of other overlapping activities, all of which have in common the processing and transmission of patterns (keystrokes, images, voice, etc.), predicated on the binary system.

The great Leibniz (1646–1716), who attributed quasi-mystical properties to the binary system, may have had a truly prophetic premonition of the place that the system was destined to occupy in the advancement of civilization. He was derided by a condescending Simon Marquis de Laplace, who wrote:

Leibnitz saw in his binary arithmetic the image of Creation.... He imagined that Unity represented God, and Zero the Void; that the Supreme Being drew all beings from the void, just as unity and zero express all numbers in his system of numeration. This conception was so pleasing to Leibnitz that he communicated it to the Jesuit Grimaldi, president of the Chinese tribunal for mathematics, in the hope that this emblem of creation would convert the Emperor of China, who was very fond of the Sciences. I mention this merely to show how the prejudices of childhood may cloud the vision even of the greatest men.

---

[21] Donald E. Knuth, *The Art of Computer Programming* (Reading, Mass.: Addison-Wesley, 1969).

[22] See Bart Kosko, *Fuzzy Thinking* (New York: Hyperion, 1993); see also Arturo Sangalli, *The Importance of Being Fuzzy* (Princeton, N.J.: Princeton University Press, 1998), and Daniel McNeil and Paul Freiberger, *Fuzzy Logic* (New York: Touchstone, 1993), as well as the seminal writings of Professor Lotfi Zadeh, inventor of the Fuzzy concept.

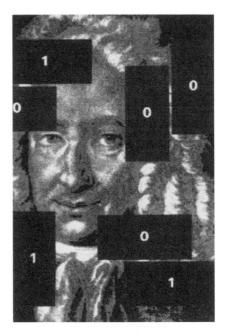

Leibniz, constellated with zeros and ones symbolizing the binary system,
as he appeared in a Siemens advertisement. Courtesy Siemens Corp.

Though Tobias Dantzig may justifiably mourn that *"what was once
hailed as a monument to monotheism ended in the bowels of a robot,"*[23]
modern-day computers, by embodying the triumph of the binary system, have indeed vindicated Leibniz. If we could count the zeros and
ones that are presently circulating every second around the world on
the Internet alone, they would no doubt far outnumber all of the remaining eight digits of the Arabic decimal alphabet ever used.

[23]Tobias Dantzig, *Number, the Language of Science*, 4th ed. (Garden City, N.Y.:
Doubleday, 1954).

CHAPTER 2

Positional Number Systems

Denaria enum ex institute hominum, non ex necessitate
naturae ut vulgus arbitratur, et sane satis inepte, posita est.
(The decimal system has been established, somewhat
foolishly to be sure, according to man's customs, not
from a natural necessity as most people would think.)
(Pascal)

This chapter is devoted to the study of positional number representation. It opens with an analysis of the division algorithm, upon
which the system is based. Following a brief introduction to codes
and their algebraic representations, the notion of mixed base is introduced, followed by the study of infinite number representations, periodic and nonperiodic. The reader should be warned that this chapter
constitutes a slight jump up in mathematical sophistication from the
preceding chapter. The only required background, however, is elementary algebra.

THE DIVISION ALGORITHM

Given arbitrary numbers 24.5 and 7.2, respectively referred to as
the *dividend* and *divisor*, we may write the following:

$$24.5 = 1 \times 7.2 + 17.3,$$
$$24.5 = 2 \times 7.2 + 10.1,$$
$$24.5 = 3 \times 7.2 + 2.9,$$
$$24.5 < 4 \times 7.2$$

Inasmuch as 3 is the largest nonnegative integer whose product by
7.2 does not exceed 24.5, it is called the *quotient*. The number 2.9

is the *remainder*, or *residue*, and is, by construction, smaller than the divisor.

If we take a close look at the division of 7.2 into 24.5, which we seem to perform naturally, and all at once, we discover that it really consists of a sequence of steps resembling the following:

1. Put $q = 1$.
2. Multiply q by 7.2
3. Compute $r = 24.5 - (q \times 7.2)$
4. Is $r \geq 7.2$?
5. If the answer to step 4 is YES, then
   i. Go to location called QUOTIENT and replace its previous content by q.
   ii. Go to location called REMAINDER and replace its previous content by r.
   iii. Increase the value of q by one, and return to step 2.
6. If the answer to step 4 is NO, then
   i. Display the content of QUOTIENT.
   ii. Display the content of REMAINDER.

The reader who is unfamiliar with computer programming will have been bewildered by that process. And yet, it precisely spells out what he or she was taught to do as a child, and continues to do, paying little or no attention to the process. When using a hand-held calculator, he or she obviously pays no attention whatsoever to that process. Actually, a calculator will not yield a quotient and a remainder as such. In the previous example, it would display 3.402777..., consisting of quotient 3 to the left of the decimal point, and *mantissa* 0.402777... to its right, which approximates 29/72

The procedure just described is called the *division algorithm*. Generally, given nonnegative dividend D and positive divisor d, the division algorithm yields

$$D = dq + \rho \qquad (0 \leq \rho < d, \quad q \text{ is an integer}). \qquad (2.1)$$

To the pair (D, d) corresponds one and only one pair of *nonnegative* numbers satisfying equation (2.1), namely the quotient q and remainder r. For a proof of that statement, see Appendix 2.1.

CODES

Given the integer $m > 1$, *integral variable* δ is said to belong to *range* m if it can take on the integral values $0, 1, 2, \ldots, m - 1$, and only these values. In other words,

$$0 \leq \delta \leq m - 1. \tag{2.2}$$

(We may indifferently say that the variable *takes on*, or *assumes*, or *is assigned* these values).

Let $\delta_0, \delta_1, \gamma$ be integral variables respectively belonging to the ranges $m_0, m_1, m_0 m_1$:

$$0 \leq \delta_0 \leq m_0 - 1, \quad 0 \leq \delta_1 \leq m_1 - 1, \quad 0 \leq \gamma \leq m_0 m_1 - 1. \tag{2.3}$$

Variables δ_0 and δ_1 are *independent*, meaning that each variable may assume any value within its range, irrespective of the value assumed by the other.

Given some agreed-upon ordering (m_0, m_1) of ranges m_0, m_1, any particular ordered pair of values (δ_0, δ_1) is called a *configuration*. There are obviously $m_0 m_1$ distinct configurations. For example, with $m_0 = 3$ and $m_1 = 2$, we have $0 \leq \delta_0 \leq 2$, $0 \leq \delta_1 \leq 1$, and $0 \leq \gamma \leq 5$. A one-to-one correspondence, or *mapping*, may be arbitrarily defined between the set of configurations (δ_0, δ_1), which are $m_0 m_1$ in number, and the set of values assignable to γ, which are also $m_0 m_1$ in number. The mapping defines a *code*, and any configuration (δ_0, δ_1) is a *representation* of the corresponding value of γ within that code. Codes may be defined in tabular form, as in Table 2.1a, which shows five different mappings out of the $6! = 720$ different possibilities corresponding to $m_0 = 3$, $m_1 = 2$.

If we regard γ as a function of independent variables δ_0, δ_1, it is possible to uniquely express γ as a (multivariate) polynomial in these variables for any randomly constructed code. Suffice it, for the moment, to indicate the simplest polynomial expressions of γ corresponding to the codes of Table 2.1a; these are shown in Table 2.1b. Observe the relative simplicity of the polynomials corresponding to codes C, D, and E. They are of the first degree, meaning that they contain no power greater than one in any one of the variables, nor do

TABLE 2.1a
A sampling of codes for $m_0 = 3$, $m_1 = 2$.

	Code									
	A		B		C		D		E	
γ	δ_1	δ_0	δ_1	δ_0	δ_1	δ_0	δ_1	δ_0	δ_1	δ_0
0	0	0	0	0	1	2	0	0	0	0
1	1	2	1	1	0	2	1	0	0	1
2	0	2	0	1	1	1	0	1	0	2
3	0	1	1	2	0	1	1	1	1	0
4	1	1	0	2	1	0	0	2	1	1
5	1	0	1	0	0	0	1	2	1	2

TABLE 2.1b
Polynomials corresponding to Table 2.1a.

Code A	$\gamma = 5\delta_0 - 2\delta_0^2 + 5\delta_1 - 5\delta_0\delta_1 + \delta_0^2\delta_1$
Code B	$\gamma = 2\delta_0 + 5\delta_1 - 9\delta_0\delta_1 + 3\delta_0^2\delta_1$
Code C	$\gamma = 5 - 2\delta_0 - \delta_1$
Code D	$\gamma = 2\delta_0 + \delta_1$
Code E	$\gamma = \delta_0 + 3\delta_1$

they contain products of the variables, such as $\delta_0\delta_1$. Additionally, neither D nor E contains a constant, whereas C does. D and E are said to be minimal codes. Clearly, two and only two minimal codes may be defined for two independent variables.

Returning to equation (2.1) and substituting integers γ for dividend D, m_0 for divisor d, and δ_0 for remainder ρ, we may write

$$\gamma = \delta_0 + m_0\delta_1, \quad \text{with} \quad 0 \leq \delta_0 \leq m_0 - 1. \qquad (2.4)$$

Making $0 \leq \delta_1 \leq m_1 - 1$ implies that $0 \leq \gamma \leq m_0 m_1 - 1$, and conversely. Generally then, given ranges m_0, m_1 and variables δ_0,

δ_1, γ satisfying (2.3), it follows that for every value of γ there exists one and only one configuration (δ_0, δ_1) satisfying equation (2.4), and conversely.

Moving on, consider ranges m_0, m_1, m_2 and independent variables $\delta_0, \delta_1, \delta_2$ respectively belonging to these ranges. Let γ' be a variable belonging to range $m_1 m_2$:

$$0 \le \gamma' \le m_1 m_2. \tag{2.5a}$$

To each value of γ' thus corresponds one and only one configuration (δ_1, δ_2) satisfying the following equation, and conversely,

$$\gamma' = \delta_1 + m_1 \delta_2. \tag{2.5b}$$

Let γ be a variable belonging to range $m_0 m_1 m_2$. We may write

$$0 \le \gamma \le m_0(m_1 m_2) - 1. \tag{2.5c}$$

To each value of γ corresponds one and only one configuration (δ_0, γ') satisfying the following equation:

$$\gamma = \delta_0 + m_0 \gamma' = \delta_0 + m_0(\delta_1 + m_1 \delta_2)$$
$$= \delta_0 + m_0 \delta_1 + m_0 m_1 \delta_2. \tag{2.6}$$

To each value of γ thus corresponds one and only one configuration $(\delta_0, \delta_1, \delta_2)$ satisfying equation (2.6), and conversely.

Introducing the coefficients $\pi_0 = 1$, $\pi_1 = m_0$, $\pi_2 = m_0 m_1$, referred to as *weights*, we get

$$\gamma = \pi_0 \delta_0 + \pi_1 \delta_1 + \pi_2 \delta_2. \tag{2.7}$$

Putting $m_0 = 3$, $m_1 = 2$, and $m_2 = 4$, the curious reader will observe that there actually exist $3! = 6$ polynomials that correspond to minimal codes. These are shown in Table 2.2. Having adopted a specific ordering of ranges m_0, m_1, m_2, consistent with their indices, polynomial a in Table 2.2, which corresponds to equation (2.7), is the only polynomial that respects this ordering. It is said to be *regular*, and constitutes the foundation upon which *positional number systems* are built, such as the decimal and binary systems. Returning to Table 2.1b, we observe that E is also the only regular code in the lot. Code D may be minimal, but it is not regular.

TABLE 2.2

Six minimal codes for $m_0 = 3$, $m_1 = 2$, $m_2 = 4$.

a	$\gamma = \delta_0 + 3\delta_1 + 6\delta_2$
b	$\gamma = 2\delta_0 + \delta_1 + 6\delta_2$
c	$\gamma = \delta_0 + 12\delta_1 + 3\delta_2$
d	$\gamma = 4\delta_0 + 12\delta_1 + \delta_2$
e	$\gamma = 8\delta_0 + \delta_1 + 2\delta_2$
f	$\gamma = 8\delta_0 + 4\delta_1 + \delta_2$

where $m_0 m_1 = 6$, $m_0 m_2 = 12$, $m_1 m_2 = 8$

MIXED-BASE POSITIONAL SYSTEMS

Base b is defined as the ordered sequence of integers:

$$b = (m_0, m_1, m_2, m_3, \ldots, m_i, \ldots, m_{n-1})$$
$$m_i > 1 \text{ for all } i. \tag{2.8}$$

It is said to be of length n, and integers $m_0, m_1, m_2, \ldots, m_i, \ldots, m_{n-1}$ are referred to as the *base ranges*, or *base radices* of base b.

Any particular assignment, within their corresponding ranges, of integral values to the variables $(\delta_0, \delta_1, \delta_2, \ldots, \delta_i, \ldots, \delta_{n-1})$ is referred to as a *conformable configuration*. In other words, a configuration is said to be *conformable* to base b if

$$0 \leq \delta_i \leq m_i - 1 \quad \text{for every } i. \tag{2.9}$$

To each index i is assigned *weight* π_i, expressed as

$$\pi_i = m_0 m_1 m_2 \ldots m_{i-1}, \text{ with } \pi_0 = 1. \tag{2.10}$$

Following the discussion of the division algorithm, it follows that to any value of integer γ belonging to range π_n, in other words, such that $0 \leq \gamma \leq \pi_n - 1$, corresponds one and only one conformable configuration $(\delta_0, \delta_1, \delta_2, \ldots, \delta_i, \ldots, \delta_{n-1})$ satisfying the following equation, and conversely.

$$\gamma = \delta_0 + m_0 \left(\delta_1 + m_1 \left(\delta_2 + m_2 \left(\delta_3 + m_3 \left(\delta_4 + \cdots + m_{n-2}\delta_{n-1}\right)\right)\right)\right)$$
$$= \pi_0 \delta_0 + \pi_1 \delta_1 + \pi_2 \delta_2 + \pi_3 \delta_3 + \cdots + \pi_{n-1}\delta_{n-1}. \qquad (2.11)$$

The parentheses are intended to provide a clue to the proof, which may be done by induction over n. By virtue of that one-to-one correspondence, the sequence $(\delta_0, \delta_1, \delta_2, \ldots, \delta_i, \ldots, \delta_{n-1})$ is said to be the *representation* of γ, base b.

That representation will be written *from right to left*, consistent with Arabic numeration. Number representations will be enclosed in parentheses, followed by a symbol identifying the base. To alleviate inconsistencies, number representations as well as base range sequences will be written from right to left, with a dot at the far right. Integer δ_i will be referred to as the ith digit of γ or the digit of *rank i*, base b, and written $(\delta_i^\gamma)_b$ or simply δ_i^γ when the base b is clear from the context. In other words,

$$\gamma = \left(\delta_{n-1}^\gamma, \ldots, \delta_i^\gamma, \ldots, \delta_2^\gamma, \delta_1^\gamma, \delta_0^\gamma.\right)_b = \sum_{i=0}^{n-1} \delta_i^\gamma \pi_i. \qquad (2.12)$$

When $m_0 = m_1 = m_2 = \cdots m_i \cdots = m$, we are in the presence of a *uniform base m*, or *uniform radix-m*, system, for which $\pi_i = m^i$; thus,

$$\gamma = \left(\delta_{n-1}^\gamma, \ldots, \delta_i^\gamma, \ldots, \delta_2^\gamma, \delta_1^\gamma, \delta_0^\gamma.\right)_m = \sum_{i=0}^{n-1} \delta_i^\gamma m^i. \qquad (2.13)$$

Otherwise, the system is referred to as a *mixed-base*, or *mixed-radix*, system.

Commas between integers may or may not be used within the actual number representations. They will be systematically used, however, in defining the base. When an arabic number is not enclosed in parentheses, however, the dot will usually be omitted. For example,

$$(1\ 0\ 1.)_2 = (1\ 2.)_3 = 5;$$

$$(2\ 1\ 1.)_a = 15, \text{ with } a = (4, 3, 2.);$$

$$(1\ 3\ 1.)_b = 15, \text{ with } b = (3, 4, 2.);$$

$$(1\ 0\ 0\ 1\ 2\ 3\ 3\ 2.)_M = 2\,165\,012, \text{ with}$$

$$M = (6, 10, 6, 10, 6, 10, 6, 10.)$$

where M represents a hypothetical Mesopotamian system.

Table 2.3 displays the correspondence between γ, which is shown in base 10 in bold characters, with its representations in bases 2, 3, and 5. Base 10 is the familiar Hindu-Arabic decimal system, and base 2 is the no less familiar binary system.

Innumerable other bases were proposed over the centuries, the most notable of which were the Mesopotamian sexagesimal and Mayan vigesimal systems. Buffon advocated the use of the *duodecimal* system, whose base 12 is divisible by 2, 3, 4, and 6, a considerable virtue in his opinion, and duodecimal fanatics went so far as to publish tables of logarithms in that system! Other systems, though not positional, were also found to have particular merits, such as the binomial, the Fibonacci, Shannon's balanced decimal system, etc.

Clearly, one may add any number of zeros to the left without changing the value of the number being represented. The leftmost nonzero digit is called the *highest significant digit*, and every digit to its right is significant. If s is the number of significant digits of integer γ, in other words, if $s - 1$ is the rank of its highest significant digit, then

$$\pi_{s-1} \le \gamma \le \pi_s - 1 \quad \text{or} \quad \pi_{s-1} - 1 < \gamma < \pi_s \qquad (2.14)$$

For example, any number γ with three significant decimal digits is such that

$$100 \le \gamma \le 999 \quad \text{or} \quad 99 < \gamma < 1000.$$

TABLE 2.3
Uniform bases **2**, **3**, and **5**.

Decimal	Base 5	Base 3	Base 2
0	0	0	0
1	1	1	1
2	2	2	10
3	3	10	11
4	4	11	100
5	10	12	101
6	11	20	110
7	12	21	111
8	13	22	1000
9	14	100	1001
10	20	101	1010
11	21	102	1011
12	22	110	1100
13	23	111	1101
14	24	112	1110
15	30	120	1111
16	31	121	10000
17	32	122	10001
18	33	200	10010
19	34	201	10011
20	40	202	10100
21	41	210	10101
22	42	211	10110
23	43	212	10111
24	44	220	11000

Similarly, any number γ with four significant binary digits, or *bits*, is such that

$$(1\ 0\ 0\ 0.)_2 \leq \gamma \leq (1\ 1\ 1\ 1.)_2 \qquad \text{or}$$

$$(1\ 1\ 1.)_2 < \gamma < (1\ 0\ 0\ 0\ 0.)_2$$

that is,

$$8 \leq \gamma \leq 15 \qquad \text{or} \qquad 7 < \gamma < 16$$

Examples abound in the literature of references to "binary numbers," or "decimal numbers." Obviously, these attributes cannot be conferred upon the numbers themselves, but upon their representations. Such expressions as "the largest conformable integer of a given length" will also be encountered. For the sake of simplicity, we shall nonetheless use linguistic shortcuts that consist of extending to the number itself the attributes of its representation, and vice versa. Another shortcut has consisted of writing

$$\gamma = (1\ 0\ 1.)_2 = (1\ 2.)_3 = 5$$

What is being stated here is that the integer whose decimal representation is 5, also has representations $(1\ 0\ 1.)_2$ in the binary system and $(1\ 2.)_3$ in the ternary system. All four representations, including the symbol γ itself, correspond to the same integer, an abstract mathematical entity, and are equivalent. A similar shortcut will consist of using the word "mantissa," upon discussing fractional numbers, to indifferently denote the fractional part of a number or its *representation*, probably to the dismay of mathematical purists. Be that as it may, we invite the reader to maintain the intellectual distinction between a number and its representations.

The largest conformable integer (LCI) of length n corresponds to $\delta_i = m_i - 1$ for every i:

$$\sum_{i=0}^{n-1} (m_i - 1)\pi_i = \sum_{i=0}^{n-1} m_i \pi_i - \sum_{i=0}^{n-1} \pi_i = \pi_n - 1. \qquad (2.15)$$

For codes a through f in Table 2.2, $\pi_3 = 24$, and the LCI is 23. For a uniform base m, we also get

$$\sum_{i=0}^{n-1} m^i = \frac{m^n - 1}{m - 1}. \tag{2.16}$$

For example,

$$1 + 10 + 100 + \cdots + 10^{n-1} = (10^n - 1)/9.$$

$$1 + 2 + 4 + 8 + \cdots + 2^{n-1} = 2^n - 1.$$

Upon closer examination, we discover that a base of length n allows the representation not only of integers $0, 1, 2, \ldots \pi_n - 1$, that is, a total of π_n integers, but of twice as many. Indeed, if only the first n base radices $(m_0, m_1, m_2, \ldots, m_{n-1})$ of a base are known, weights $\pi_0, \pi_1, \pi_2, \ldots, \pi_{n-1}$ may be calculated, *as well as* $\pi_n = m_{n-1}\pi_{n-1}$. Regardless of what m_n may be, and it is necessarily larger than 1, a conformable number of length $n + 1$ can always be created for every conformable number of length n, by adding the digit 1 in the otherwise empty position of rank $i = n$. Strictly speaking, then, a base of length n allows the representation of $2\pi_n$ integers of length $n + 1$, namely, 0 to $2\pi_n - 1$.

Finding the Digits of an Integer

Given base b, and integer γ expressed in some different base, say, decimal, several algorithms allow the determination of the integer's digits, base b. The two methods described below (as well as every other method for that matter) follow directly from the division algorithm.

METHOD I

1. Divide γ by m_0. The quotient is q_1 and the remainder δ_0^γ
2. Divide q_1 by m_1. The quotient is q_2 and the remainder δ_1^γ
3. Divide q_2 by m_2. The quotient is q_3 and the remainder δ_2^γ

$\cdots \ \cdots$

s. Divide q_{s-1} by m_{s-1}. The quotient is 0 and the remainder δ_{s-1}.

Example 2.1

$$b = (5, 4, 3, 2 .): \pi_0 = 1, \pi_1 = 2, \pi_2 = 6, \pi_3 = 24, \pi_4 = 120.$$

i. $\gamma = 115 = 2 \times 57 + 1$
$57 = 3 \times 19 + 0$
$19 = 4 \times 4 + 3$
$4 = 5 \times 0 + 4 \qquad 115 = (4\ 3\ 0\ 1 .)_b$

ii. $\gamma = 24 = 2 \times 12 + 0$
$12 = 3 \times 4 + 0$
$4 = 4 \times 1 + 0$
$1 = 5 \times 0 + 1 \qquad 24 = (1\ 0\ 0\ 0 .)_b = \pi_3$

iii. $\gamma = 119 = 2 \times 59 + 1$
$59 = 3 \times 19 + 2$
$19 = 4 \times 4 + 3$
$4 = 5 \times 0 + 4 \qquad 119 = (4\ 3\ 2\ 1 .)_b = \pi_4 - 1,$
$\qquad\qquad\qquad\qquad\quad (119 \text{ is the LCI of length } 4).$

METHOD 2

The base conversion algorithm described above may be reexpressed step by step as follows, where symbol $[A]$ represents the integral part of number A:

1. $\gamma = m_0 \left[\dfrac{\gamma}{m_0} \right] + \delta_0^\gamma$

2. $\left[\dfrac{\gamma}{m_0} \right] = m_1 \left[\dfrac{\gamma}{m_0 m_1} \right] + \delta_1^\gamma$

3. $\left[\dfrac{\gamma}{m_0 m_1} \right] = m_2 \left[\dfrac{\gamma}{m_0 m_1 m_2} \right] + \delta_2^\gamma$

$\cdots$

In general,

$$\delta_i^\gamma = \left[\frac{\gamma}{\pi_i} \right] - m_i \left[\frac{\gamma}{\pi_{i+1}} \right] = \left[m_i \frac{\gamma}{\pi_{i+1}} \right] - m_i \left[\frac{\gamma}{\pi_{i+1}} \right]. \qquad (2.17a)$$

A theorem of number theory states that

$$[mA] - m[A] = ([mA] \bmod m), \qquad (2.17b)$$

where $(A \bmod m)$ denotes the remainder of the division of A by m. Thus

$$\delta_I^\gamma = \left(\left[\frac{\gamma}{\pi_i} \right] \bmod m_i \right). \qquad (2.17c)$$

Example 2.2

i. To express $\gamma = 115$ in the factorial base,

$$\delta_0^\gamma = \left(\left[\frac{115}{1} \right] \bmod 2 \right) = 1$$

$$\delta_1^\gamma = \left(\left[\frac{115}{2} \right] \bmod 3 \right) = 0$$

$$\delta_2^\gamma = \left(\left[\frac{115}{6} \right] \bmod 4 \right) = 3$$

$$\delta_3^\gamma = \left(\left[\frac{115}{24} \right] \bmod 5 \right) = 4.$$

ii. To convert $\gamma = 315$ from decimal to binary,

$$\delta_0^\gamma = \left(\left[\frac{315}{1} \right] \bmod 2 \right) = 1$$

$$\delta_1^\gamma = \left(\left[\frac{315}{2} \right] \bmod 2 \right) = 1$$

$$\delta_2^\gamma = \left(\left[\frac{315}{4} \right] \bmod 2 \right) = 0$$

$$\delta_3^\gamma = \left(\left[\frac{315}{8} \right] \bmod 2 \right) = 1$$

$$\delta_4^\gamma = \left(\left[\frac{315}{16} \right] \bmod 2 \right) = 1$$

$$\delta_5^\gamma = \left(\left[\frac{315}{32} \right] \bmod 2 \right) = 1$$

$$\delta_6^\gamma = \left(\left[\frac{315}{64} \right] \bmod 2 \right) = 0$$

$$\delta_7^\gamma = \left(\left[\frac{315}{128} \right] \mathrm{mod}\, 2 \right) = 0$$

$$\delta_8^\gamma = \left(\left[\frac{315}{256} \right] \mathrm{mod}\, 2 \right) = 1.$$

Since $315 < 2^i$ for all powers i higher than 8, δ_8^γ is the highest significant digit, and the required number is $(1\,0\,0\,1\,1\,1\,0\,1\,1.)_2$

The first base conversion method is *iterative*. In order to calculate digit δ_i^γ, one must calculate digits δ_j^γ for every index j from 0 to $i - 1$. The second method has the advantage of providing an explicit formulation for δ_i^γ, allowing the calculation of every digit on its own, inasmuch, of course, as the calculation of a number's integral part may be regarded as deriving from a truly explicit formulation not requiring iteration, notwithstanding the essentially algorithmic nature of the division process itself.

Addition

To arithmetic operations on numbers correspond rules for manipulating their representations. These rules, which in the case of decimal arithmetic are learned early in life, become second nature to everyone. What really accounts for the universal triumph of the positional decimal system over all previous systems is the invention of simple recursive procedures for performing arithmetic. These procedures are open ended; in other words, they are valid regardless of the length of numbers being manipulated. Additionally, because positional number systems are linear, arithmetic operations are also independent of digit rank. Let us begin with an example.

Let a and b be two decimal digits. If both a and b are equal to 9, their sum is equal to 18. That sum consists of two digits, namely, a *partial sum*, 8, and a *carry* of 1. There is no other way of writing the number 18 in the decimal system. If we add any two decimal digits $0 \le a,\ b \le 9$, their sum obviously lies between 0 and 18, meaning that its decimal representation consists at most of two decimal digits, namely, a *partial sum s* no larger than 8, and a *carry c* no larger than 1.

Consider now the sum $(1 + 9 + 9) = 19$. In this case, also, the carry is 1. Consequently, if we add $(1 + a + b)$, the partial sum may not

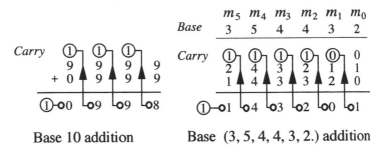

Base 10 addition **Base (3, 5, 4, 4, 3, 2.) addition**

Figure 2.1. The carry never exceeds one when adding two numbers.

exceed 9, and the carry may not exceed 1. Adding any two numbers consisting of more than one digit, no carry larger than 1 may thus originate anywhere in the process, as in Figure 2.1.

That reasoning obviously applies to any base, as in Figure 2.2a. For every rank i, the algorithm allows the determination of the partial sum s_i and the carry c_i, corresponding to the sum $(c_{i-1} + \delta_i^\alpha + \delta_i^\beta)$. Equation (2.17c) yields in this case

$$s_i = \left((c_{i-1} + \delta_i^\alpha + \delta_i^\beta) \bmod m_i\right) \quad \text{and}$$

$$c_i = \left(\left[\frac{c_{i-1} + \delta_i^\alpha + \delta_i^\beta}{m_i}\right] \bmod m_{i+1}\right), \tag{2.17d}$$

and since no carry may exceed 1 when adding only two numbers, we can simply write

$$c_i = \left[\frac{c_{i-1} + \delta_i^\alpha + \delta_i^\beta}{m_i}\right]. \tag{2.17e}$$

Example 2.3

To add $\alpha = (2\ 4\ 3\ 2\ 1\ 1.)_b = 1431$ and $\beta = (1\ 4\ 3\ 3\ 2\ 0.)_b = 958$, where base $b = (3, 5, 4, 4, 3, 2.)$, follow the steps shown in Figure 2.2a. The reader may verify that $(1\ 1\ 4\ 3\ 2\ 0\ 1.)_b = 2389$. Using the traditional addition format, we have the calculation shown in Figure 2.2b. Observe that the sum requires seven digits, whereas base length is 6. That situation is perfectly legitimate, given that the

i	m_i	$\delta_i^a \; \delta_i^b$		Carry	Sum
0	2	$1+0$	$=1 = (0\,1.)_2$	$carry = 0$	$sum = 1$
1	3	$1+2$	$=3 = (1\,0.)_3$	$carry = 1$	$sum = 0$
2	4	►$1+2+3$	$=6 = (1\,2.)_4$	$carry = 1$	$sum = 2$
3	4	►$1+3+3$	$=7 = (1\,3.)_4$	$carry = 1$	$sum = 3$
4	5	►$1+4+4$	$=9 = (1\,4.)_5$	$carry = 1$	$sum = 4$
5	3	►$1+2+1$	$=4 = (1\,1.)_3$	$carry = 1$	$sum = 1$
					1

Figure 2.2a. $(243211.)_b + (143320.)_b = (1143201.)_b$ where $b = (3, 5, 4, 4, 3, 2.)$

	Base	3	5	4	4	3	2
	Carry	1	1	1	1	0	0
α		2	4	3	2	1	1
β		1	4	3	3	2	0
	1	1	4	3	2	0	1

Figure 2.2b. Same addition as in 2.2a, put in classical form.

last carry equals 1 and is therefore conformable to any extension by one position of the base length.

Example 2.4

To add $14 = (1110.)_2$ to $5 = (101.)_2$, follow the procedure shown in Figure 2.2c.

	Carry	1	1	0	0	0
	14	0	1	1	1	0
	5	0	0	1	0	1
	1	0	0	1	1	

Figure 2.2c. $(1110.)_2 + (101.)_2 = (10011.)_2$

Uniform-Base Multiplication

Uniform-base multiplication of two numbers is done in very much the same manner as traditional base 10 multiplication, as illustrated by the following examples.

Example 2.5

To multiply $315 \times 5 = 1575$ in binary, use the calculation shown in Figure 2.3a. Nonsignificant zeros were deliberately added to the multiplicand and the product for the sake of completing the number of their respective binary digits to twelve. These particular twelve-bit numbers, including their nonsignificant zeros, were chosen to illustrate *cyclic* numbers, which will be studied later. The reader will observe that if the product and multiplicand digits are arranged in the form of closed twelve-bit loops, the two resulting loops are identical, albeit rotated with respect to one another. The next example involves a base 63 cyclic number.

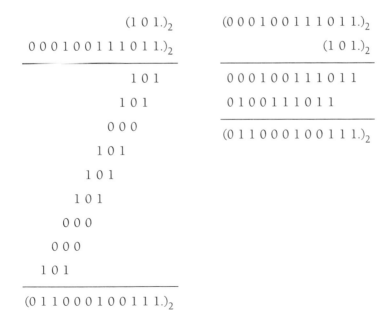

Figure 2.3a. Multiplying a cyclic binary number.

$(12 \ 37 \ 50 \ 25.)_{63}$		$(12 \ 37 \ 50 \ 25.)_{63}$		$(12 \ 37 \ 50 \ 25.)_{63}$	
	$\times \ \ 2$		$\times \ \ 3$		$\times \ \ 4$
	50		1 12		1 37
1 37		2 24		3 11	
1 11		1 48		2 22	
24		36		48	
$(25 \ 12 \ 37 \ 50.)_{63}$		$(37 \ 50 \ 25 \ 12.)_{63}$		$(50 \ 25 \ 12 \ 37.)_{63}$	

Figure 2.3b. Multiplying a cyclic number, base 63.

Example 2.6

Figure 2.3b shows multiplication in base 63 for

$$3\,150\,592 \times 2 = 6\,301\,184,$$

$$3\,150\,592 \times 3 = 9\,451\,776,$$

$$3\,150\,592 \times 4 = 12\,602\,368.$$

Figure 2.3c shows the multiplication of 4325 by 2164, as performed by the Arabs. The reader may easily decipher how it works.

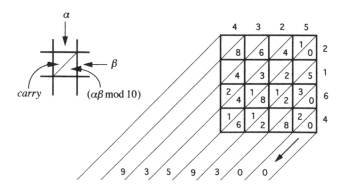

Figure 2.3c. Arabic multiplication. *Source*: Ibn el Hatem al-Makdassy, *Al-Hawi f'il hissab*, transcribed by R. el Salihy and K. al-Monshidawy of Baghdad University's Faculty of Engineering (Baghdad: National Library, 1988, p. 55).

Mixed-Base Multiplication

Mixed-base multiplication is more complex, as illustrated by an example using the factorial base $f = (\ldots 5, 4, 3, 2\,.)$. Its *companion* bases are defined as follows:

$$f_0 = f = (\ldots 5, 4, 3, 2\,.), \qquad f_1 = (\ldots 6, 5, 4, 3\,.)$$
$$f_2 = (\ldots 7, 6, 5, 4\,.), \qquad f_3 = (\ldots 8, 7, 6, 5\,.)$$

$\ldots$

Example 2.7

Multiply $N = 89 = (3\ 2\ 2\ 1\,.)_f$ by 7.
We first perform the following operations:

$$7 \times \delta_0^N = 7 \times 1 = 7 = (1\ 0\ 1\,.)_{f_0}$$

$$7 \times \delta_1^N = 7 \times 2 = 14 = (1\ 0\ 2\,.)_{f_1}$$

$$7 \times \delta_2^N = 7 \times 2 = 14 = (3\ 2\,.)_{f_2}$$

$$7 \times \delta_3^N = 7 \times 3 = 21 = (4\ 1\,.)_{f_3}$$

Then we add, remembering that the modulus changes from one column to the next, as shown in Figure 2.3d.

m_i	6	5	4	3	2
Carry	1	0	0	0	0
			1	0	1
		1	0	2	
	3	2			
	4	1			
Sum	(5	0	3	2	1.)$_f = 623$

Figure 2.3d. Multiplying in the factorial base.

CONSTRUCTION 1: A PARALLEL ADDER

Electrical engineers and computer designers are adept at constructing elaborate assemblies of interconnected tiny devices in a manner that constitutes the physical embodiment of some prescribed function. Each device has a number of input ports and one output port, and is said to *realize* a particular elementary transfer function. Arithmetic machines are good examples, as illustrated by Figure 2.4a. All units are compatible, in the sense that one device's output port may be connected to another device's input port, and the numerical values assigned to the signals traveling between ports belong to prescribed well-defined sets, discrete or continuous. Figure 2.4b shows the construction of a *parallel* adder, using the adder of Figure 2.4a.

The earliest arithmetic machine, the Pascaline, was invented by Blaise Pascal (Figure 2.5). Clever arrangements of gears allowed the generation of each partial sum, as it propagated the carry from one position to the next-higher-rank position.

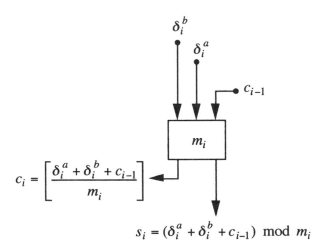

$$c_i = \left[\frac{\delta_i^a + \delta_i^b + c_{i-1}}{m_i} \right]$$

$$s_i = (\delta_i^a + \delta_i^b + c_{i-1}) \bmod m_i$$

Figure 2.4a. An elementary full adder (which produces both the partial sum and the carry).

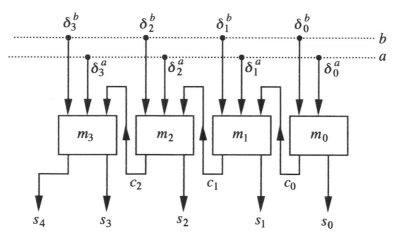

Figure 2.4b. A four-digit parallel adder.

Figure 2.5. Blaise Pascal operating his Pascaline, a mechanical adding and subtracting machine of his own invention. From an engraving in *Machines et inventions approuvées par l'Académie Royale des Sciences,* published by Jean-Gauffin Gallon in Paris in 1735. Reproduced by permission of Sigvard Strandh, *A History of the Machine* (Göteborg, Sweden: AB Nordbok, 1979).

CONSTRUCTION 2: A DIGITAL-TO-ANALOG CONVERTER

The basic building block we shall use is schematically represented by Figure 2.6a. If input signals a and b are respectively applied to the input ports, output signal c is given by $c = a + mb$. Constant m is the device's *gain*.

$$b \rightarrow \boxed{m} \rightarrow c = a + mb$$

Figure 2.6a. The basic building block.

The construction of Figure 2.6b consists of a cascade of devices identical to the basic building block of Figure 2.6a, whose gains are adjusted as shown. Input signal δ_i^γ, measured on an appropriate scale, is allowed to take on the values $0, 1, 2, \ldots, m-1$. Similarly, the meter shown on the right is calibrated to measure, on the same scale, signals ranging from 0 to π_{n-1}.

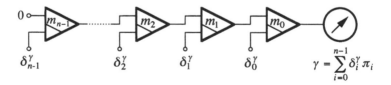

$$0 \rightarrow \boxed{m_{n-1}} \cdots \boxed{m_2} \boxed{m_1} \boxed{m_0} \rightarrow \bigcirc$$

$$\delta_{n-1}^\gamma \qquad \delta_2^\gamma \qquad \delta_1^\gamma \qquad \delta_0^\gamma \qquad \gamma = \sum_{i=0}^{n-1} \delta_i^\gamma \pi_i$$

Figure 2.6b. A digital-to-analog converter.

The device converts a number, expressed in digital form, to an analog magnitude, measured on an appropriate scale. The above example is extremely simplified, and real-life digital-to-analog converters are obviously more elaborate.

CONSTRUCTION 3: A REVERSIBLE
BINARY-TO-ANALOG CONVERTER

The following mathematical amusement is offered as an illustration of binary numeration. It may be realized in actuality, and serve as an educational toy or a science fair project. Consider the pulley of Figure 2.7a, whose axis is allowed to move longitudinally as the string that is wrapped around it is pulled from either side. Spring S is intended only to provide a reaction to the rope's pull, and ensure the construction's stability. In what follows, we shall concern ourselves only with longitudinal displacements and rotations, not with forces. If x and y denote the longitudinal rope displacements and z denotes the resulting displacement of the pulley's axis in the direction shown, then $z = (x + y)/2$.

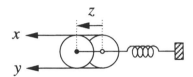

Figure 2.7a. The pulley.

Figure 2.7b represents a device that is intended to, say, position a read-write head H on the surface of a ruler of length L, which is divided into sixteen identical segments. The left-hand side input devices consist of little pistons, freely sliding inside their respective cylinders, also of length L. A piston may be pulled or pushed to either extremity of its cylinder, but may not come to rest anywhere in between. The input signal consists of four binary digits, or bits, namely

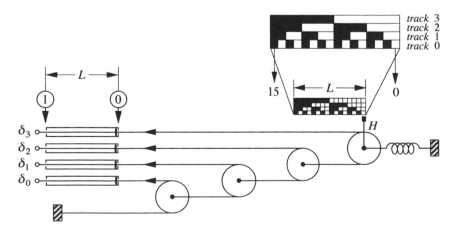

Figure 2.7b. A binary-to-analog converter.

δ_0, δ_1, δ_2, δ_3. Input $\delta_i = 0$ signals that piston i must be moved to the right of its cylinder, and $\delta_i = 1$ signals that the piston must be moved to the left. The reader may verify that when δ_i goes from 0 to 1, the head moves from a white box on track i to the black box immediately to its left, through a displacement equal to $L \times (2^i/16)$. The total head displacement d upon the ruler is thus given by

$$d = \frac{L}{16}(\delta_0 + 2\delta_1 + 4\delta_2 + 8\delta_3). \qquad (2.18a)$$

Analog-to-digital conversion is difficult to achieve using the transducers of Figure 2.6b, and will not be dealt with here. On the other hand, the machine of Figure 2.7b is perfectly reversible. If springs are placed at the left-hand extremity, one for each piston, the analog input at H is converted to its binary representation, where digits $\delta_0, \delta_1, \delta_2, \delta_3$ are now the outputs.

POSITIONAL REPRESENTATION OF FRACTIONAL NUMBERS

A nonnegative number is said to be fractional when it is smaller than 1. The word "fractional" is somewhat ambiguous, as it may convey the notion that the number at hand can be represented by a fraction, such as 2/5—in other words, that it is rational. Such is not necessarily the case, and "fractional" refers to both rational and irrational numbers, smaller than 1. A real number consists of an integral part and a fractional part, either of which may be zero. If the fractional part is rational, the number itself may be represented by a fraction, meaning that it is rational. If the fractional part is irrational, the number is similarly irrational. An integer (including zero) is a rational number.

In the preceding sections, we have used the symbol γ to denote an integer. In what follows, the symbol μ will denote a number in general, consisting of an integral part, which may be zero, and a fractional part, which may also be zero. We shall also sometimes use the symbol ν to denote a fractional number. We now address the problem of representing fractional numbers within the framework of the positional system.

Consider base $b_0 = (5, 4, 2, 4, 3, 2\,.\,)$ of length 6, and the number $\mu = (2\ 3\ 1\ 0\ 2\ 1\,.\,)_{b_0} = 557$, also of length 6, which is conformable

TABLE 2.4a
The number $\mu = (2\ 3\ |\ 0\ 2\ 1\,.)_{b_0}$, base $b_0 = (5, 4, 2, 4, 3, 2\,.)$.

i	6	5	4	3	2	1	0
m_i		5	4	2	4	3	2
π_i	960	192	48	24	6	2	1
δ_i^{μ}		2	3	1	0	2	1

$$\mu = (1 \times 1) + (2 \times 2) + (6 \times 0) + (24 \times 1) + (48 \times 3) + (192 \times 2) = 557.$$

to base b_0. Table 2.4a shows the succession of ranges and weights corresponding to that base, along with the digits of μ.

We now decide that in moving the radix dot from the immediate right of index $i = 0$ to that of index $i = 3$, for example, the configuration $\mu' = (2\ 3\ 1.\ 0\ 2\ 1)_{b_3}$ is generated, along with its companion base $b_3 = (5, 4, 2.\ 4, 3, 2)$. We shall stipulate that

$$\mu' = (2\ 3\ 1.\ 0\ 2\ 1)_{b_3} = \frac{(2\ 3\ 1\ 0\ 2\ 1\,.)_{b_0}}{\pi_3}.$$

We now define new weights, as shown in Table 2.4b, where

$$\pi'_0 = 1 \qquad \pi'_1 = m'_0 \qquad \pi'_2 = m'_0 m'_1 \cdots$$

$$\pi'_{-1} = \frac{1}{m'_{-1}} \qquad \pi'_{-2} = \frac{1}{m'_{-1} m'_{-2}} \cdots$$

Then we multiply each digit $\delta_k^{\mu'}$ by its corresponding weight π'_k and add

$$\mu' = (1 \times 1) + (2 \times 3) + (8 \times 2) = 23 \qquad \text{left of the dot}$$

$$+ (1/4 \times 0) + (1/12 \times 2) + (1/24 \times 1) = 5/24 \quad \text{right of dot}$$

$$= 23 5/24 = 557/24.$$

That result is consistent with our initial stipulation, and suggests that a base consisting of a sequence of integers with a radix dot somewhere within may be viewed as the juxtaposition of two distinct and

TABLE 2.4b
The number $\mu' = (2\ 3\ 1.\ 0\ 2\ 1)_{b_3}$, base $b_3 = (5,4,2\ .\ 4,3,2)$.

k	3	2	1	0 .	−1	−2	−3
m'_k		5	4	2 .	4	3	2
π'_k	40	8	2	1 .	¼	1⁄12	1⁄24
δ^μ_k		2	3	1 .	0	2	1

independent bases on either side of the dot. The left-hand base allows the representation of the integral part of a number (23 in the example), sometimes called the *argument*, and the right-hand base allows the representation of the *mantissa*, which corresponds to its fractional part (5/24 in the example). These bases need not bear any relationship to each other, any more than a radix need bear any relationship to any other radix, provided that they are all larger than 1. Generally, given *reference* base b_0 and the number μ,

$$b_0 = (m_{L-1}, \ldots, m_i, \ldots, m_1, m_0 .$$
$$m_{-1}, m_{-2}, \ldots, m_{-i}, \ldots, m_{-R}), \quad (2.18b)$$

and

$$\mu = (\delta^\mu_{L-1}, \ldots, \delta^\mu_i, \ldots, \delta^\mu_1, \delta^\mu_0 .$$
$$\delta^\mu_{-1}, \delta^\mu_{-2}, \ldots, \delta^\mu_{-i}, \ldots, \delta^\mu_{-R}). \quad (2.18c)$$

Putting

$$\pi_0 = 1, \quad \pi_1 = m_0, \quad \pi_2 = m_0 m_1, \quad \pi_3 = m_0 m_1 m_2, \ldots,$$

$$\pi_{-1} = \frac{1}{m_{-1}}, \quad \pi_{-2} = \frac{1}{m_{-1} m_{-2}}, \quad \pi_{-3} = \frac{1}{m_{-1} m_{-2} m_{-3}}, \ldots,$$

$$(2.18d)$$

we get

$$\mu = \sum_{i=-R}^{L-1} \pi_i \delta_i^{\mu}. \tag{2.19}$$

For any index i, $-R \leq i \leq L-1$, companion base b_i is obtained from base b_0 by placing the radix dot to the immediate right of index i. That definition applies to both positive and negative values of i; thus,

$$b_i = (m_{I-1}, \ldots, m_i \bullet m_{i\ 1}, \ldots, m_0, m_{-1}, \ldots, m_{-R}) \tag{2.20a}$$

$$b_{-i} = (m_{L-1}, \ldots, m_0, m_{-1}, \ldots, m_{-i} \bullet m_{-i-1}, \ldots, m_{-R}) \tag{2.20b}$$

Returning to the example of Table 2.4b,

$$(2\ 3\ 1\ 0\ 2\ 1.)_{b_0}/\pi_3 = (2\ 3\ 1.\ 0\ 2\ 1)_{b_3} = 557/24,$$

$$\text{with } b_3 = (5, 4, 2\,.\,4, 3, 2)$$

$$(2\ 3\ 1\ 0\ 2\ 1.)_{b_0}/\pi_5 = (2.\ 3\ 1\ 0\ 2\ 1)_{b_5} = 557/192,$$

$$\text{with } b_5 = (5\,.\,4, 2, 4, 3, 2)$$

For uniform-base systems, the companion base is the base itself, regardless of index i. For example,

$$1234567/100 = 12345\,.\,67$$

$$(110100110.)_2/8 = (110100\,.\,110)_2$$

A number is said to be *representable* or *expressible* within a given *finite* base if there exists a conformable configuration for the number in that base. We shall later see that any number is expressible in any positional system of infinite length.

The largest conformable mantissa of length R is

$$\sum_{i=-1}^{-R}(m_i-1)\pi_i = \frac{m_{-1}-1}{m_{-1}} + \frac{m_{-2}-1}{m_{-1}m_{-2}} + \cdots + \frac{m_{-R}-1}{m_{-1}m_{-2}m_{-3}\cdots m_{-R}}$$

$$= 1 + \left(-\frac{1}{m_{-1}} + \frac{1}{m_{-1}}\right) + \left(-\frac{1}{m_{-1}m_{-2}} + \frac{1}{m_{-1}m_{-2}}\right) + \cdots$$

$$= 1 - \frac{1}{m_{-1}m_{-2}m_{-3}\cdots m_{-R}} = 1 - \pi_{-R}. \qquad (2.21)$$

The $1/\pi_{-R}$ expressible mantissas are multiples of π_{-R}, namely,

$$0, \quad \pi_{-R}, \quad 2\pi_{-R}, \quad 3\pi_{-R}, \quad \ldots, \quad (1-\pi_{-R}).$$

The minimum nonzero number expressible by a mantissa of length R is π_{-R}, and the numbers that can be expressed by a mantissa of length R and an argument of length L are all multiples of π_{-R}. They are shown in Table 2.5a in the form of an array of π_L lines and $1/\pi_{-R}$ columns. To each line corresponds one value of the argument, and to each column corresponds one value of the mantissa.

Adding π_{-R} to the last number of a row resets the mantissa to 0, and carries 1 left of the dot, incrementing the argument by 1. The resulting number is the first number on the following line. The mantissa's cycle then repeats itself, until the largest conformable number (LCN) of total length $L + R$ is reached, which is $(\pi_L - \pi_{-R})$.

TABLE 2.5a
The numbers 0 to $(\pi_L - \pi_{-R})$ in increments of π_{-R}.

0	π_{-R}	$2\pi_{-R}$	$\ldots$	$1 - \pi_{-R}$
1	$1 + \pi_{-R}$	$1 + 2\pi_{-R}$	$\ldots$	$2 - \pi_{-R}$
2	$2 + \pi_{-R}$	$2 + 2\pi_{-R}$	$\ldots$	$3 - \pi_{-R}$
$\ldots$	$\ldots$	$\ldots$	$\ldots$	$\ldots$
$(\pi_L - 1)$	$((\pi_L - 1) + \pi_{-R})$	$((\pi_L - 1) + 2\pi_{-R})$	$\ldots$	$(\pi_L - \pi_{-R})$

Following the pattern of Table 2.5a, the successive numbers that are expressible without changing the dot's location are shown in Table 2.5b for two different settings of the radix dot, where SCN is the smallest nonzero conformable number, LCM is the largest conformable mantissa, LCI is the largest conformable integer, and LCN is the largest conformable number. The total number of expressible numbers is $40 \times 24 = 960$ in the left-hand example, and $20 \times 48 = 960$ in the right-hand example.

Similarly, Table 2.5c shows the conformable number in the factorial base, where the total number of such numbers is $24 \times 24 = 576$, in increments of 1/24.

TABLE 2.5b

The smallest conformable number > 0 (SCN), the largest conformable mantissa (LCM), the largest conformable integer (LCI), and the largest conformable number (LCN) for bases $(5, 4, 2 . 4, 3, 2)$ and $(5, 4 . 2, 4, 3, 2)$.

	Base	Base
	$5\ 4\ 2\ .\ 4\ 3\ 2$	$5\ 4\ .\ 2\ 4\ 3\ 2$
	$0\ 0\ 0\ .\ 0\ 0\ 0\ =\ 0$	$0\ 0\ .\ 0\ 0\ 0\ 0\ =\ 0$
SCN > 0	$0\ 0\ 0\ .\ 0\ 0\ 1\ =\ \frac{1}{24}$	$0\ 0\ .\ 0\ 0\ 0\ 1\ =\ \frac{1}{48}$
	$0\ 0\ 0\ .\ 0\ 1\ 0\ =\ \frac{2}{24}$	$0\ 0\ .\ 0\ 0\ 1\ 0\ =\ \frac{2}{48}$
	$0\ 0\ 0\ .\ 0\ 1\ 1\ =\ \frac{3}{24}$	$0\ 0\ .\ 0\ 0\ 1\ 1\ =\ \frac{3}{48}$
	. . .	
LCM	$0\ 0\ 0\ .\ 3\ 2\ 1\ =\ \frac{23}{24}$	$0\ 0\ .\ 1\ 3\ 2\ 1\ =\ \frac{47}{48}$
	$0\ 0\ 1\ .\ 0\ 0\ 0\ =\ \frac{24}{24} = 1$	$0\ 1\ .\ 0\ 0\ 0\ 0\ =\ \frac{48}{48} = 1$
	$0\ 0\ 1\ .\ 0\ 0\ 1\ =\ 1\frac{1}{24}$	$0\ 1\ .\ 0\ 0\ 0\ 1\ =\ 1\frac{1}{48}$
	. . .	
	$0\ 0\ 1\ .\ 3\ 2\ 1\ =\ 1\frac{23}{24}$	$0\ 1\ .\ 1\ 3\ 2\ 1\ =\ 1\frac{47}{48}$
	$0\ 0\ 2\ .\ 0\ 0\ 0\ =\ 2$	$0\ 2\ .\ 0\ 0\ 0\ 0\ =\ 2$
	. . .	
LCI	$4\ 3\ 1\ .\ 0\ 0\ 0\ =\ 39$	$4\ 3\ .\ 0\ 0\ 0\ 0\ =\ 19$
	$4\ 3\ 1\ .\ 0\ 0\ 1\ =\ 39\frac{1}{24}$	$4\ 3\ .\ 0\ 0\ 0\ 1\ =\ 19\frac{1}{48}$
	. . .	
LCN	$4\ 3\ 1\ .\ 3\ 2\ 1\ =\ 39\frac{23}{24}$	$4\ 3\ .\ 1\ 3\ 2\ 1\ =\ 19\frac{47}{48}$

TABLE 2.5c
Same exercise as in Table 2.5b, with the
reflected factorial base (4 3 2 . 2 3 4).

	Base
	4 3 2. 2 3 4
	0 0 0 . 0 0 0 = 0
SCN > 0	0 0 0 . 0 0 1 = $\frac{1}{24}$
	0 0 0 . 0 0 2 = $\frac{2}{24}$
	0 0 0 . 0 0 3 = $\frac{3}{24}$
	0 0 0 . 0 1 0 = $\frac{4}{24}$
	. . .
LCM	0 0 0 . 1 2 3 = $\frac{23}{24}$
	0 0 1 . 0 0 0 = 1
	0 0 1 . 0 0 1 = $1\frac{1}{24}$
	. . .
	0 0 1 . 1 2 3 = $1\frac{23}{24}$
	0 1 0 . 0 0 0 = 2
	. . .
LCI	3 2 1 . 0 0 0 = 23
	3 2 1 . 0 0 1 = $23\frac{1}{24}$
	. . .
LCN	3 2 1 . 1 2 3 = $23\frac{23}{24}$

For any given dot location within a base, the number of expressible numbers of length $(L + R)$ is π_L/π_{-R}, in other words, the product P of all base radices. That number is invariant. For every dot
location within the base, we may write P distinct number representations. The precision of the base b number system therefore depends
not on the size π_{-R} of the smallest increment, but on the total number P of expressible numbers. The position of the base dot is immate-
rial in terms of precision, and *floating-point* systems are based on that
idea. They consist in stating the entire array of length $L + R$, without base dot, accompanied by a statement of the value of R. In deci-

mal notation, for example, the number 12345678E-3 is interpreted as $12345678 \times 10^{-3} = 12345.678$. The number $R = 3$ is referred to as the exponent.

Going to Infinity

Open-ended bases may be created, which are infinitely extensible left and right of the radix point. One may imagine an infinity of such open-ended bases, with corresponding rules for base-radix generation. A symmetrical, or reflected, base is one for which

$$m_{i-1} = m_{-i} ; \quad \text{thus } \pi_{-i} = \frac{1}{\pi_i}, \quad \text{with } \pi_0 = 1. \quad (2.22)$$

If the base is uniform and symmetrical, $m_i = m$ and $\pi_i = m^i$ for every i, positive, zero, or negative. The most classical of all uniform reflected systems is the decimal system. The other widespread uniform reflected system is, of course, the binary system.

The largest conformable mantissa (LCM) of length R represents the number $(1 - \pi_{-R})$. As R tends to infinity, π_{-R} tends to zero, and $(1 - \pi_{-R})$ tends to one. It follows from (2.21) that

$$\sum_{i=-1}^{-\infty} (m_i - 1)\pi_i = 1, \quad (2.23)$$

meaning that the series corresponding to the LCM converges to one in any positional system, whether uniform or mixed. An infinite number of infinitely close numbers, in other words, the *continuum of real numbers*, may thus be represented, ranging from zero to one. In decimal representation, the largest conformable mantissa of infinite length is $.999\ldots \to 1$, signifying that the mantissa converges to one. We shall write $.999\ldots = 1$. Similarly,

$$1 = (.1111\ldots)_2 = (.2222\ldots)_3 = (.3333\ldots)_4 \cdots \quad (2.24)$$

An interesting geometric metaphor for the statement

$$1 = (.1111\ldots)_2 = \frac{1}{2} + \frac{1}{4} + \frac{1}{8} + \frac{1}{16} + \cdots \quad (2.25)$$

is provided by Figure 2.8a, where the area of right triangle ABC is $\frac{1}{2}(\sqrt{2})^2 = 1$. Drawing BD perpendicular to AC divides the triangle

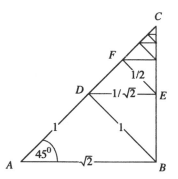

Figure 2.8a. Geometric metaphor for $1 = (.1111\ldots)_2$.

into identical right triangles ABD and BCD, each of area 1/2. Similarly BCD is divided into BDE and CDE, each of area 1/4. The process may be extended to infinity, verifying equation (2.25).

Similarly, a metaphor for the statement

$$1 = (.3333\ldots)_4 \quad \text{or} \quad 1 = \frac{3}{4} + \frac{3}{16} + \frac{3}{64} + \frac{3}{256} \cdots \quad (2.26)$$

is provided by Figure 2.8b. The large equilateral triangle of step 1 is of area 1. It is subdivided into four identical equilateral triangles, each of area 1/4. The area of the black portion, which excludes the central triangle, is thus 3/4. That triangle is now subdivided into four identical triangles, each of area $(1/4)/4 = 1/16$. The total black area at step 2 is now $3/4 + 3/16$. Similarly, the black area at step 3 is $3/4 + 3/16 + 3/64$. If the process is extended to infinity, the central little triangle tends to vanish, verifying equation (2.26).

step 1   step 2   step 3

Figure 2.8b. Geometric metaphor for $1 = (.3333\ldots)_4$.

An interesting base is the *reflected factorial* base

$$f = (\ldots 6, 5, 4, 3, 2 \cdot 2, 3, 4, 5, 6, \ldots) \qquad (2.27)$$

In that base,

$$(.123456\ldots)_f = 1, \qquad (2.28)$$

and corresponding to Euler's number e is the representation

$$e = 2 + \frac{1}{2!} + \frac{1}{3!} + \frac{1}{4!} = (1\,0\,.\,1\,1\,1\,1\ldots)_f \qquad (2.29a)$$

That constitutes a striking example of a transcendental number representation by a periodic mantissa. (Obviously, the base itself is not periodic). Similarly, we may write

$$\cosh 1 = (1\,.\,1\,0\,1\,0\,1\,0\,1\ldots)_f \qquad (2.29b)$$

$$\sinh 1 = (1\,.\,0\,1\,0\,1\,0\,1\,0\ldots)_f \qquad (2.29c)$$

and

$$\cosh(1) + \sinh(1) = (1\,0\,.\,1\,1\,1\,1\,1\ldots)_f = e. \qquad (2.29d)$$

As a practical exercise, we may numerically "verify," using a computer, the convergence to one, the largest conformable mantissa for the factorial base. With S_j denoting the largest conformable mantissa of length j, base f, the left panel of Table 2.6 shows the successive values of S_j for increasing values of j, namely, $1/2!$, $1/2! + 2/3!$, $1/2! + 2/3! + 3/4!$, and so on.

With sixteen-digit precision, S_j rapidly converges to 1. With infinite precision, if such a thing were to exist, it would take an infinite number of steps to "reach" 1. As an additional exercise, we may verify that the LCM converges to 1 for any randomly constructed base. The algorithm is the following:

Pick a maximum range M larger than 2.
Pick integer δ_1 at random among integers $1, 2, \ldots, M - 1$,
    and put

$$\pi_1 = \delta_1 + 1 \qquad S_1 = \frac{\delta_1}{\pi_1}.$$

TABLE 2.6
Convergence of two infinite series.

Factorial base		Random Base (M = 4)	
δ_j	S_j	δ_j	S_j
1	.5	1	.5
2	.75	2	.8333333333333334
3	4.916666666666667	3	.9583333333333334
4	.9791666666666667	3	.9895833333333334
5	.9958333333333333	3	.9973958333333334
6	.9993055555555556	1	.9986979166666667
7	.9999007936507937	2	.9995659722222222
8	.9999875992063492	2	.9998553240740741
9	.9999986221340388	1	.9999276620370370
10	.9999998622134039	3	.9999819155092593
11	.9999999874739458	3	.9999954788773148
12	.9999999989561622	2	.9999984929591050
13	.9999999999197048	3	.9999996232397762
14	.9999999999942646	1	.9999998116198881
15	.9999999999996177	3	.9999999529049721
16	.9999999999999761	3	.9999999882262430
17	.9999999999999986	2	.9999999960754143
18	.9999999999999999	2	.9999999986918048
19		3	.9999999996729512
20		3	.9999999999182378
21		3	.9999999999795595
22		3	.9999999999948899
23		2	.9999999999982966
24		2	.9999999999994322
25		1	.9999999999997161
26		3	.9999999999999290
27		2	.9999999999999764
28		2	.9999999999999921
29		1	.9999999999999961
30		3	.9999999999999990
31		2	.9999999999999997
32		2	.9999999999999999

Pick integer δ_2 at random among integers $1, 2, \ldots, M - 1$, and put

$$\pi_2 = \pi_1(\delta_2 + 1) \qquad S_2 = S_1 + \frac{\delta_2}{\pi_2}.$$

Repeat, putting

$$\pi_j = \pi_{j-1}(\delta_j + 1) \qquad S_j = S_{j-1} + \frac{\delta_j}{\pi_j}$$

The algorithm generates a random succession of ranges between 2 and M, and S_j is the LCM at every stage. In that case, also, S_j rapidly converges to 1 (right panel of Table 2.6).

How Precise Is a Mantissa?

It must be understood that, contrary to an integer's representation, which is finite in length, a number's mantissa is essentially of infinite length. Whereas each new position to the left of the radix dot brings with it increasing finite weights, a fractional number is the sum of the terms of a convergent infinite series, with each term finite, though evanescent. Whereas a mantissa is by definition infinite in length, any real-life number representation, for practical reasons, is necessarily finite. Most modern personal computers, for example, will typically handle sequences comprising no more than sixteen decimal digits.

As we shall later see, rational number representations are either finite or periodic (when the base itself is periodic, as in the decimal and binary bases). A periodic number, say, .123412341234..., may be represented by the "compressed" expression .$\underline{1234}$. Actually, a finite mantissa, such as .25, may be regarded as a compression of the periodic representation .250000... followed by an infinite number of zeros, or of .249999... followed by an infinite number of 9s, that is, .25 = .25$\underline{0}$ = .24$\underline{9}$. Similarly, $(. 1\ 0\ 1\ 1)_2 = (. 1\ 0\ 1\ 0\ \underline{1})_2$. In other words, finite-length expressions always exist for rational numbers. Nonetheless, these expressions, periodic or not, compressed or not, may be too long for practical purposes, and have to be abridged in some way or another.

As for irrational numbers, their representation is always infinite, never periodic. No shortcut therefore exists, and the numbers always have to be truncated somewhere, whenever they are put to practical use. Truncation is usually accompanied by rounding. For example, π may be rounded to 3.14, 3.142, 3.1416, 3.14159, etc. Truncation, with or without rounding, necessarily entails an uncertainty, whose relative magnitude depends on the base used and where truncation occurs.

Whereas any algorithm for the generation of an integer's digits must be relentlessly pursued until the integer's significant numbers are exhausted (otherwise the result is meaningless), an algorithm for the generation of a fractional number's mantissa is bounded by physical limits. If the algorithm, within the imposed limits, discovers a rational number's full-length representation, finite or periodic, that is fine, and a statement to that effect is made, using the equality sign. Otherwise, a statement is made in the form lower limit $\leq \mu <$ upper limit. When the number is irrational, we need to replace the symbol $\leq$ by $<$. For example:

i. $\mu = .25$ is as an *exact* full-length expression, representing the rational number 1/4. That is precisely the meaning of the equality sign.

ii. $\mu = .33333$ is an *exact* full-length finite expression, representing rational number 33333/100000, whereas $x = .333\ldots = .\overline{3} = 1/3$, which is also an exact expression.

iii. $3.14159 < \pi < 3.14160$,

or $\qquad 314\,159 < 100\,000\pi < 314\,160.$ $\qquad$ (2.30a)

The latter expressions involve a truncated unrounded decimal representation. With only six digits, it may or may not be sufficiently precise for the purpose at hand. Similarly,

$$(1.0\ 2\ 1\ 4\ 4)_f < \sqrt{2} < (1.0\ 2\ 1\ 4\ 5)_f$$

which means that

$$1018 < 720\sqrt{2} < 1019. \qquad (2.30b)$$

It follows from equation (2.21) that

$$\sum_{i=-(R+1)}^{-\infty} (m_i - 1)\pi_i = \pi_{-R}, \qquad (2.31)$$

meaning that the LCM of infinite length whose initial R digits right of the radix dot are replaced by zeros is equal to π_{-R}. The decimal base thus yields.

$$1 = \frac{9}{10} + \frac{9}{100} + \frac{9}{1000} + \frac{1}{1000}. \tag{2.32a}$$

Similarly, the factorial base yields

$$1 = \frac{1}{2} + \frac{2}{6} + \frac{3}{24} + \frac{4}{120} + \frac{5}{720} + \frac{1}{720}, \tag{2.32b}$$

and the binary base yields

$$1 = \frac{1}{2} + \frac{1}{4} + \frac{1}{8} + \frac{1}{16} + \frac{1}{32} + \frac{1}{64} + \frac{1}{64}. \tag{2.32c}$$

This expression corresponds to the Horus Eye myth (Plate 5).

Generally, if the number ν_R corresponds to the first R digits of a truncated mantissa, the uncertainty that remains regarding the mantissa's true value is π_{-R}. Thus,

$$\nu_R \leq \nu < \nu_R + 1/\pi_{-R}. \tag{2.33}$$

For a reflected base, $\pi_{-R} = 1/\pi_R$, a number μ_R, whose mantissa is of length R, corresponds to

$$\mu_R \leq \mu < \mu_R + \pi_{-R}, \tag{2.34a}$$

or

$$(\mu_R \pi_R) \leq \mu \pi_R < (\mu_R \pi_R) + 1. \tag{2.34b}$$

Observe that inequalities (2.30a) and (2.30b) are consistent with (2.34b). The ancient Greeks discovered that

$$3\frac{10}{71} < \pi < 3\frac{1}{7} \tag{2.35}$$

and

$$\frac{239}{169} < \sqrt{2} < \frac{99}{70}. \tag{2.36}$$

Statement (2.35), which was provided by Archimedes, may be written as

$$1\,561 < \pi \times 497 < 1\,562.$$

Statement (2.36), which was provided by Theodorus of Cyrene, may be written as

$$16730 < \sqrt{2} \times 11830 < 16731.$$

Both statements are also consistent with (2.34b).

Finding the Digits of a Fractional Number

Consider the base $b = (\ldots m_2, m_1, m_0 . m_{-1}, m_{-2}, \ldots)$. It is required to determine the first R digits of the fractional number ν, which is given within another base, or in some other form, such as $\nu = (3/7)$, or $\nu = \sqrt{2} - 1$. We have already examined two methods for the generation of an integer's argument. The first of the two methods may be transposed to the mantissa as follows:

Multiply ν by m_{-1}. This results in a number consisting of integral part $[\nu m_{-1}]$ and fractional part ν_1.
Putting $[\nu m_{-1}] = \delta_{-1}$, we may write

$$\nu \times m_{-1} = \delta_{-1} + \nu_1 \qquad 0 \le \nu_1 < 1$$
$$\nu_1 \times m_{-2} = \delta_{-2} + \nu_2 \qquad 0 \le \nu_2 < 1$$
$$\nu_2 \times m_{-3} = \delta_{-3} + \nu_3 \qquad 0 \le \nu_3 < 1$$

$$\cdots \qquad\qquad\qquad \cdots$$

$$\nu_{R-1} \times m_{-R} = \delta_{-R} + \nu_R \qquad 0 \le \nu_R < 1.$$

Consistent with (2.33), this eventually yields

$$\left(.\delta^{\nu}_{-1}, \delta^{\nu}_{-2}, \ldots, \delta^{\nu}_{-R}\right)_b \le \nu < \left(.\delta^{\nu}_{-1}, \delta^{\nu}_{-2}, \ldots, \delta^{\nu}_{-R}\right)_b + \pi_{-R}$$

$$(2.37)$$

Example 2.8

i. Express 1222/9900 in decimal form, as shown in Figure 2.9. Observe that the fractional part on the right side of the fourth line is identical to that on the left side of the third line. The loop thus formed generates a never-ending iterative process, which results in a periodic mantissa. That mantissa is $.1234343434\ldots = .12\underline{34}$

$$\begin{aligned} 10\,(1222/9900) &= 1 + (232/990)\\ 10\,(232/990) &= 2 + (34/99)\\ \rightarrow 10\,(34/99) &= 3 + (43/99)\\ 10\,(43/99) &= 4 + (34/99) \end{aligned}$$

Figure 2.9. The periodic representation of 1222/9900, base 10.

ii. Express 3/7 within the factorial base:

$$\begin{aligned} 2\,(3/7) &= 0 + 6/7\\ 3\,(6/7) &= 2 + 4/7\\ 4\,(4/7) &= 2 + 2/7\\ 5\,(2/7) &= 1 + 3/7\\ 6\,(3/7) &= 2 + 4/7\\ 7\,(4/7) &= 4 + 0. \end{aligned}$$

Thus $3/7 = (.022124)_f = 2/6 + 2/24 + 1/120 + 2/720 + 4/5040$. It is clear from this example that to rational numbers always correspond finite mantissas in the factorial system.

iii. Express $(\sqrt{2} - 1)$ within the factorial base:

$$\begin{aligned} 2(\sqrt{2} - 1) &= 0 + (2\sqrt{2} - 2)\\ 3(2\sqrt{2} - 2) &= 2 + (6\sqrt{2} - 8)\\ 4(6\sqrt{2} - 8) &= 1 + (24\sqrt{2} - 33)\\ 5(24\sqrt{2} - 33) &= 4 + (120\sqrt{2} - 169)\\ 6(120\sqrt{2} - 169) &= 4 + (720\sqrt{2} - 1018)\\ 7(720\sqrt{2} - 1018) &= 1 + (5040\sqrt{2} - 7127), \end{aligned}$$

which gives, with six-digit precision,

$$(.021441)_f < \sqrt{2} - 1 < (.021442)_f,$$

or $\quad 2087/5040 < \sqrt{2} - 1 < 2088/5040. \qquad (2.38a)$

How did we arrive at the above statements, in the absence of any other positional representation of $\sqrt{2} - 1$? The answer is given in Appendix 2.2.

As an additional exercise, the reader may verify that

$$(1.011010100000100111100)_2$$

$$< \sqrt{2} < (1.011010100000100111101)_2 \qquad (2.38b)$$

Similar algorithms are generally available for the so-called *algebraic* numbers, to which we shall return. Among these are *surds*, such as $\sqrt{2}$. For numbers such as π or Euler's number e, base conversion is always possible, however, from one available approximate representation to another, or from known representations of the number, other than positional, as we shall see in the chapter on cleavages.

Finding the Digits of a Real Number

A real number consists of an (integral) argument and a mantissa. The problem now consists of determining both components. In the preceding discussion, we have shown how to calculate the digits of a (nonnegative) number's integral argument, and those of its mantissa. The attentive reader will have understood that the methods shown are fundamentally the same, and may be lumped into a single over-arching method of great simplicity. Indeed, equation (2.17c) applies to all values of i, negative as well as nonnegative, and may thus be rewritten as

$$\delta_i^\mu = \left\{ \left[\frac{\mu}{\pi_i} \right] \bmod m_i \right\}. \qquad (2.39)$$

Given real number μ, whose argument $[\mu]$ does not exceed $\pi_L - 1$, and whose required mantissa is limited to R digits, we have

$$\left(\delta_{L-1}^\nu, \ldots, \delta_1^\nu, \delta_0^\nu . \delta_{-1}^\nu, \delta_{-2}^\nu, \ldots, \delta_{-R}^\nu \right)_b$$

$$\leq \mu < \left(\delta_{L-1}^\nu, \ldots, \delta_1^\nu, \delta_0^\nu . \delta_{-1}^\nu, \delta_{-2}^\nu, \ldots, \delta_{-R}^\nu \right)_b + \pi_{-R} \qquad (2.40)$$

For example, $\sqrt{2}$ expressed in the factorial base yields

$$(1.\,0\,2\,1\,4\,4\,1)_f < \quad \sqrt{2} \quad < (1.\,0\,2\,1\,4\,4\,2)_f$$

$$(1\,0\,2\,1\,4\,4\,1.)_{f_{-6}} < 5040\sqrt{2} < (1\,0\,2\,1\,4\,4\,2.)_{f_{-6}}$$

(2.41)

That corresponds to the procedure shown in Table 2.7.

TABLE 2.7
Finding the digits of $\sqrt{2}$ in the factorial base.

i	m_i	π_i	$[\mu/\pi_i]$	$\left(\delta_i^{\sqrt{2}}\right)_f$
1	3	2	0	0
0	2	1	1	1
−1	2	1/2	2	0
−2	3	1/6	8	2
−3	4	1/24	33	1
−4	5	1/120	169	4
−5	6	1/720	1018	4
−6	7	1/5040	7127	1

PERIODIC BASES

An important class of open-ended systems is that of reflected periodic, or cyclical bases, such as $b = (\ldots 5,\ 2,\ 3,\ 5,\ 2,\ 3\,.\ 3,\ 2,\ 5,\ 3,\ 2,\ 5,\ \ldots)$, for which we may use the abridged representation $(5, 2, 3 . 3, 2, 5)$. Both bases shown are said to be of cycle or period $\tau = 3$. The product of the base radices within one full cycle is $\pi_\tau = 30$, and $\pi_{-\tau} = 1/30$.

Whether a base is periodic or not, mantissas may or may not be periodic within that base. For example, the mantissa of Euler's number e is periodic when represented in the factorial base, which is not periodic. Such is also the case for cosh 1 and sinh 1. An underlining scheme similar to that used for periodic bases will also be used for periodic mantissas. Thus $e = (10.\underline{1})_f$, cosh $1 = (1.\underline{10})_f$,

$\sinh 1 = (1.\underline{01})_f$. For simplicity, we shall sometimes use the expression "periodic number," instead of the lengthy "a number whose representation is periodic."

In what follows, we shall study the properties of periodic number representations only when the base is reflected periodic. The following examples illustrate a general property of periodic mantissas.

Example 2.9

    i. Base $b_0 = (4, 3.\underline{3}, 4)$, $\pi_\tau = 12$.

$$\mu = (.\underline{1\,2}) = (1/3 + 2/12) + (1/36 + 2/144)$$

$$+ (1/432 + 2/1728) + \cdots$$

$$= (6/12) + (6/144) + (6/1728) + \cdots$$

$$= 6\big(1/12 + 1/(12)^2 + 1/(12)^3 + \cdots\big).$$

In other words,

$$\mu = (.1\,2)_{b_0} \times \pi_\tau \left(\left(\frac{1}{\pi_\tau}\right) + \left(\frac{1}{\pi_\tau}\right)^2 + \left(\frac{1}{\pi_\tau}\right)^3 + \cdots \right),$$

and given that for any number $k > 1$,

$$\sum_{i=1}^{\infty} \frac{1}{k^i} = \frac{1}{k-1}, \tag{2.42}$$

we get

$$\mu = \frac{(.12)_{b_0} \pi_\tau}{\pi_\tau - 1} = \frac{(12.)_{b_{-\tau}}}{\pi_\tau - 1} = \frac{6}{11},$$

with $b_{-\tau} = (\dots, 4, 3, 3, 4.\underline{3}, 4, 3, 4, \dots)$.

    ii. Base $b_0 = (.\underline{2}, 3, 5)$, $\pi_\tau = 30$.

$$(.\underline{1\,2\,3})_{b_0} = (1\,2\,3.)_{b_{-3}}/29 = 28/29,$$

$$(.\underline{1\,1\,2})_{b_0} = (1\,1\,2.)_{b_{-3}}/29 = 36/29,$$

$$(.\underline{1\,2\,4})_{b_0} = (1\,2\,4.)_{b_{-3}}/29 = 29/29 = 1.$$

where $b_{-3} = (\dots 5, 3, 2, 2, 3, 5.\underline{2}, 3, 5 \dots)$

Example 2.10

When the mantissa's period is not equal to that of the base, the mantissa's length taken into consideration is the least common multiple (LCM) of the two periods. Here, base $b_0 = (.2, 3)$, $\pi_7 = 6$.

$$(.\underline{1})_{b_0} = (.\underline{11})b_0 = (11.)_{b_{-2}}/5 = 4/5,$$

$$(.\underline{1\ 2\ 0\ 1})_{b_0} = (1\ 2\ 0\ 1.)_{b_{-4}}/(36 - 1) = 31/35,$$

$$(.\underline{1\ 0\ 1})_{b_0} = (.\underline{1\ 0\ 1\ 1\ 0\ 1})_{b_0}$$

$$= (1\ 0\ 1\ 1\ 0\ 1.)_{b_{-6}}/215 = 133/215.$$

Example 2.11

For uniform base m systems, $\pi_t = m^t$ for any t.

$$0.\underline{1} = 1/9 \qquad 0.\underline{12} = 12/99 \qquad 0.\underline{174} = 174/999,$$

$$(.\underline{1\ 0\ 1})_2 = (1\ 0\ 1.)_2/(2^3 - 1) = 5/7,$$

$$(.\underline{0\ 2\ 1\ 1})_3 = (2\ 1\ 1.)_3/(3^4 - 1) = 22/80 = 11/40.$$

Example 2.12

We will also encounter mantissas consisting of an initial finite-length sequence, called the *mantissa lead sequence*, or *mantissa leader*, followed by a periodic sequence of infinite length, such as $0.243333\ldots = 73/300$. With base $b = (.2, 3)$, we get

$$(.0\ 0\ \underline{1})_b = (.\underline{1\ 1})_b/6 = (4/5)/6 = 2/15,$$

$$(.0\ \underline{1})_b = (.0\ 1)_b + (.\underline{1\ 1})_b/6 = 1/6 + (4/5)/6 = 3/10,$$

$$(.1\ 2\ \underline{1\ 0\ 1})_b = (.1\ 2)_b + (.\underline{1\ 0\ 1})_b/6$$

$$= 5/6 + (133/215)/6 = 604/645.$$

The above three sequences are said to be *quasi-periodic*, or *semi-periodic*. By contrast, sequences without a leader are said to be *strictly periodic*. The adjective *periodic* thus applies to both sequence types. A finite mantissa may be regarded as quasi-periodic, consisting of a leader followed by an infinite periodic sequence of zeros. Obviously, *finite* mantissas may present a character of periodicity. The term *periodic* or *cyclical*, must, however, be understood to apply only to infinite-length mantissas. For example, $0.2525 = 2525/10000$, whereas $0.\underline{2525} = 25/99 = 2525/9999$.

Given a quasi-periodic mantissa ν consisting of a lead sequence of length s followed by an infinite periodic sequence, let t be the smallest common multiple of the mantissa's period and that of the base. It follows from the preceding discussion that any periodic mantissa is such that $\nu = A/\pi_s + B/\pi_s(\pi_t - 1)$, where A and B are integers; that is,

$$\nu = N/\pi_s(\pi_t - 1), \tag{2.43}$$

where N, s, and t are integers, and t is a multiple of τ. When the base itself is periodic, periodic mantissas thus always represent rational numbers. We will later show that the converse of that statement is also true, namely, that a rational number's mantissa is always periodic (or finite, or null) *when the base itself is periodic*, whether that base is mixed or uniform.

A Triadic (Ternary) Yardstick

Figure 2.10 shows how a triadic yardstick may be constructed for the measurement of distances between 0 and 1, in increments of 1/81. Stage 1 consists of dividing the distance between 0 and 1 into three equal segments. The leftmost segment contains the 0 mark, and extends infinitely closely to the 1/3 mark, without containing it. Similarly, the second segment contains the 1/3 mark but not the 2/3 mark, and the rightmost segment contains the 2/3 mark but not the 1 mark. Stages 2, 3, and 4 consist of subdividing, in like fashion, each segment of the previous stage. To each stage corresponds a new digit position within the mantissa, to the right of the preceding positions. Stage 1 indicates that the coordinate of A is equal to or larger than 0, but smaller than 1/3, and the coordinate of B is equal to or larger than $(.1)_3 = 1/3$, but smaller than 2/3. Stage 4 indicates that $(.0121.)_3 \leq A < (.0122)_3$ and $(.1012)_3 \leq B < (.1020)_3$. With n stages, the yardstick allows the measurement of distances in increments of $1/3^n$. The reason for the $\leq$ and $<$ signs is that every interval contains its origin, but excludes its end point.

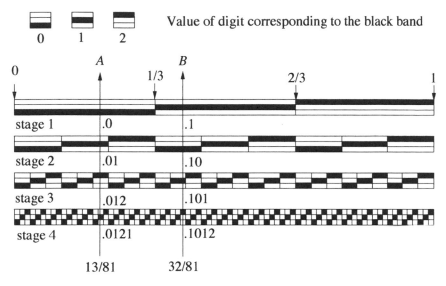

Figure 2.10. A triadic yardstick.

MARGINALIA

Unit Fractions Revisited

In what follows, we shall not attempt to rediscover the some-times obscure methods that ancient Egyptians resorted to in order to construct their unit fraction tables. As an exercise, however, we shall represent any given fraction within an ad hoc mixed-base system in such a manner that all of the number's digits are ones. That process only sometimes generates the Egyptian unit fraction expansion.

Given the fraction a/d, (with $a < d$), we proceed as prescribed under "Finding the Digits of a Fractional Number," with the difference that the exercise now consists in finding not the digits, but the *base radices* corresponding to a hypothetical mantissa made up only of ones. We may proceed as follows:

Pick the lowest integer m_{-1} such that $d < m_{-1}a < 2d$, and write

$$m_{-1}\frac{a}{d} = 1 + \frac{s_1}{d} \quad (s_1 \text{ is an integer}).$$

Pick the lowest integer m_{-2} such that $d < m_{-2}s_1 < 2d$, and write

$$m_{-2}\frac{s_1}{d} = 1 + \frac{s_2}{d} \quad (s_2 \text{ is an integer})$$

Continue this process until $d = m_{-n}s_{n-1}$. We may now write

$$a/d = (.\;\underbrace{1\ 1\ 1 \ldots 1\ 1\ 1}_{\leftarrow\; n \text{ ones} \;\rightarrow})_b, \quad b = (.m_{-1}, m_{-2}, m_{-3}, \ldots m_{-n}).$$

Example 2.13

   i. $a/d = 2/13.$
      $7 \times 2/13 = 1 + 1/13$
      $13 \times 1/13 = 1.$
      Hence $2/13 = (.11)_b$, with $b = (.7, 13)$,   or
      $2/13 = 1/7 + 1/91.$

  ii. $a/d = 7/15.$
      $3 \times 7/15 = 1 + 6/15$
      $3 \times 6/15 = 1 + 3/15$
      $5 \times 3/15 = 1.$
      Hence $7/15 = (.111)_b$, with $b = (.3, 3, 5)$,   or
      $7/15 = 1/3 + 1/9 + 1/45.$

The reader may establish that the above process always leads to a zero remainder, and thus allows the generation of finite representations. As a further exercise, let us examine what may happen when instead of picking the *lowest* integer $m_i + 1$ such that $d < m_{i+1}s_i < 2_d$, we pick the *highest* integer satisfying that condition.

$$
\begin{aligned}
2 \times 11/13 &= 1 + 9/13 \\
2 \times\ \ 9/13 &= 1 + 5/13 \\
5 \times\ \ 5/13 &= 1 + 12/13 \\
2 \times 12/13 &= 1 + 11/13
\end{aligned}
$$

Figure 2.11. Obtaining a periodic sum of unit fractions.

For example, for $a/d = 11/13$, the process shown in Figure 2.11 yields $11/13 = (.1)_b$, with $b = (.\underline{2, 2, 5, 2})$, and the reader may verify the following sequence:

1/13: $b = (.25, 2, \overline{2, 2, 5, 2})$
2/13: $b = (.12, 2, \overline{2, 5, 2})$
3/13: $b = (.8, \overline{2, 2, 5, 2})$
4/13: $b = (.6, \overline{2, 2, 5, 2})$
5/13: $b = (.5, 2, \overline{2, 2, 5, 2})$

$\cdots \quad \cdots$

10/13: $b = (.2, 3, 3, \overline{2, 2, 5, 2})$
11/13: $b = (.2, \overline{2, 5, 2})$
12/13: $b = (.\overline{2, 2, 2, 5, 2})$

We are in the presence of semi-periodic bases, which consist of a finite nonperiodic lead sequence followed by an infinite periodic sequence. Remember that the reflected factorial base representation of the transcendental number e is $(1\ 0\ .\ \underline{1})_f$

APPENDIX 2.1

To prove that given nonnegative dividend D, positive divisor d, and the equation

$$D = dq + r \quad (0 \le r < d, q \text{ is an integer}),$$

to the pair (D, d) corresponds one and only one pair of *nonnegative* numbers satisfying the equation, namely, the quotient q and remainder r. We shall reason by reductio ad absurdum, as follows:

If that statement were not true, another pair (q', r') could be found, where q' is also an integer and $0 \le r' < d$, such that

$$D = dq + r = dq' + r', \quad \text{i.e.,} \quad d(q - q') = (r' - r).$$

That would lead to the following contradictions:

1. If $q \ne q'$, the absolute value $|q - q'|$ must be an integer equal to or larger than one, which implies that the absolute value $|r - r'|$ is equal to or larger than d. In other words,

$$q \ne q' \rightarrow |q - q'| \ge 1 \rightarrow |r - r'| = d|q - q'| \ge d.$$

   That is impossible, since by definition both r' and r are nonnegative and smaller than d. Thus $q = q'$.
2. Because $q = q'$, subtracting $D = dq' + r'$ from $D = dq + r$ yields $r - r' = dq' - dq = d(q' - q) = 0$, thus $r = r'$.

APPENDIX 2.2

How is the statement $5(24\sqrt{2} - 33) = 4 + (120\sqrt{2} - 169)$ arrived at? The procedure is based on the statement $(\sqrt{2})^2 = 2$, which constitutes the definition of $\sqrt{2}$.

Consider the left side of the given equation, and put $x = 5(24\sqrt{2} - 33)$. This gives $(x + 165)^2 = (120\sqrt{2})^2 = 28800$

Now "test" $x = 1, 2, 3 \ldots$ and record the consecutive values of $(x + 165)^2$:

$$(1 + 165)^2 = 27225 < 28800$$
$$(2 + 165)^2 = 27889 < 28800$$
$$(3 + 165)^2 = 28224 < 28800$$
$$(4 + 165)^2 = 28561 < 28800$$
$$(5 + 165)^2 = 28900 > 28800$$

As 5 is the first value of x for which $(x + 165)^2$ exceeds 28800, x must be larger than 4 and smaller than 5. We may thus write $x = 4 + s = 5(24\sqrt{2} - 33)$. Hence $s = 120\sqrt{2} - 169$, and we finally get $5(24\sqrt{2} - 33) = 4 + (120\sqrt{2} - 169)$.

CHAPTER 3

Divisibility and Number Systems

This name (Theory of numbers) is given to that part of
algebra in which we ask questions about factors, the
divisibility of certain numbers by other numbers, the
possibility of expressing numbers by means of algebraic
expressions of certain kinds, and other things of that sort.
(E. C. Titchmarsh)[1]

This chapter deals with the relationship between divisibility and
positional numeration, and opens with a brief exposition of the fundamental theorem of arithmetic and congruences. Applying Euler's
theorem and a generalization thereof, it establishes the periodicity of
rational number representation within periodic bases, and derives the
means for calculating the period and lead length. Finally, it examines
some aspects of ancient problems, such as that of cyclic numbers,
among others. This chapter constitutes a substantial jump in difficulty
from the previous two chapters. It may be skipped altogether if found
difficult to read, without affecting the reader's understanding of subsequent chapters.

THE FUNDAMENTAL THEOREM OF ARITHMETIC

It is self-evident that any natural number A may be divided by 1
and itself. A and 1 are said to be *factors* of A. A factor of A other than
1 and A itself is called a *proper* factor. A prime number is defined as
a natural number that has no proper factors. In accordance with these
definitions, natural number 1, which has no proper factor, should
be regarded as prime, which it generally is *not*. That is a matter of
convention, and sometimes leads to ambiguities.

[1]E. C. Titchmarsh, *Mathematics for the General Reader* (New York: Dover, 1981),
p. 45.

A number other than 1 that is not prime is said to be *composite*, and the *fundamental theorem of arithmetic*, whose proof was first given by Gauss in his *Disquisitiones arithmeticae* in 1801, states that any composite number can be uniquely (i.e., in only one way) expressed as a product of prime factors (where each factor is allowed to occur more than once). For example $660 = 2^2 \times 3 \times 5 \times 11$ and $18900 = 2^2 \times 3^3 \times 5^2 \times 7$. That statement does not appear as such in Euclid's *Elements*, but was accepted over the centuries as self-evident, which it definitely is not.

Generally one may write $A = a^\alpha b^\beta c^\gamma \ldots$, $(a, b, c > 1)$, where a, b, c are the *prime factors* of integer A, and $\alpha, \beta, \gamma \ldots (\alpha, \beta, \gamma > 0)$ their respective *multiplicities*.

Had the integer 1 been considered a prime number, it would have been a prime factor of every natural number with arbitrary multiplicities, which is deemed undesirable.

Two numbers m and d are said to be *relatively prime* if they have no common prime factors, or, equivalently, if they have no common divisor other than 1. That is written $(m, d) = 1$, where the symbol (m, d) denotes the greatest common divisor of m and d. This is also written $m \perp d$.

If two numbers are relatively prime, there is obviously no power to which either one may be raised that will make it a multiple of the other, or, indeed, of any factor thereof. If integers m and d are not relatively prime, let d_0 be the largest factor of d that is relatively prime to m. With $d = d_0 \bar{d}$, we get

$$\left(d_0, \bar{d}\right) = \left(m, d_0\right) = 1. \tag{3.1a}$$

The smallest integer x such that m^x is a multiple of $\bar{d}$ shall be called the exponent of m over d, and denoted s. It follows that

$$\left(d, m^x\right) = d/d_0 = \bar{d} \quad \text{if and only if } x \geq s. \tag{3.1b}$$

That statement suggests the following definition:

The exponent of m over d is defined as the smallest value of exponent x for which $d/(d, m^x)$ is relatively prime to m.

When m and d are relatively prime, we shall put $s = 0$, consistent with $\bar{d} = (d, m^0) = 1$.

Example 3.1

$$m = 11 \times 2^2 \times 3 \times 5 = 660, \quad d = 7 \times 2^2 \times 3^3 \times 5^2 = 18900.$$

$$s = 3, \qquad (d, m^3) = (11^3 \times 2^6 \times 3^3 \times 5^3, 7 \times 2^2 \times 3^3 \times 5^2)$$

$$= (2^2 \times 3^3 \times 5^2).$$

hence $\quad d/(d, m^3) = 7 = d_0.$

CONGRUENCES

According to Gauss's own definition, two numbers A and B are said to be *congruent for the modulus d*, or *congruent modulo d*, when their difference is divisible by positive integer d. That relationship is expressed as

$$A \equiv B \pmod{d}. \tag{3.2a}$$

For example, $14 \equiv 39 \equiv 4 \pmod{5}$ means that $(14 - 4)$, $(39 - 4)$, and $(39 - 14)$ are all divisible by 5. If the difference between two numbers is negative, it can be made congruent to a positive number by adding a sufficiently large multiple of d. The division of integer D by integer d yields one and only one integer pair (q, ρ), where q is the quotient and ρ is the remainder, or residue, where

$$D = dq + \rho \qquad 0 \le \rho < d. \tag{3.2b}$$

We shall write

$$\rho = (D \bmod d) \tag{3.2c}$$

Whereas expression (3.2a) signifies a *relationship* between the two sides of congruence sign "$\equiv$," the term $(D \bmod d)$ shall denote a residue between 0 and $d - 1$. We may therefore write $(A \bmod a) = (B \bmod b)$, signifying that the remainder of the division of A by a is equal to the remainder of the division of B by b. For example, $(25 \bmod 3) = (10^6 \bmod 7) = (-8 \bmod 9) = 1$.

Congruence defines an *equivalence* relation. Equivalence is defined by the following four properties:

1. *Determination.* Either $A \equiv B \pmod{d}$ or $A \not\equiv B \pmod{d}$.
2. *Reflexivity.* $A \equiv A \pmod{d}$.
3. *Transitivity.* $A \equiv B$ and $B \equiv C$ jointly imply $A \equiv C$ $\pmod{d}$.
4. *Symmetry.* $A \equiv B$ implies $B \equiv A \pmod{d}$.

In what follows, we list without proof a few essential properties of congruences. The reader should be careful not to automatically regard the converse of any of the following statements as true. As a matter of fact, that would be the exception.

1. $A \equiv B$ *and* $C \equiv D$ *jointly imply*

$$A + B \equiv C + D \pmod{d};$$

$$A - B \equiv C - D \pmod{d}; \quad and$$

$$AC \equiv BD \pmod{d}.$$

   Hence,
   $A^n \equiv B^n \pmod{d}$ *for any integer n; and*
   $A \equiv B$ *implies* $Ak \equiv Bk \pmod{d}$, *where k is any integer.*
2. $A \equiv B \pmod{d}$ *implies* $A \equiv B \pmod{d'}$ *for any divisor* d' *of d.*
3. $A \equiv B \pmod{d}$ *and* $A \equiv B \pmod{c}$ *jointly imply*
   $A \equiv B \pmod{(c, d)}$, *where* (c, d) *is the least common multiple of c, d, and conversely; and*
   If c and d are relatively prime, $A \equiv B \pmod{cd}$, *and conversely.*
4. $Ak \equiv Bk \pmod{d}$ *implies* $A \equiv B \pmod{l}$, *with*
   $l = d/(k, d)$, *where* (k, d) *is the greatest common divisor of k and d. Hence* $Ak \equiv Bk \pmod{d}$ *implies* $A \equiv B \pmod{d}$ *if k is relatively prime to d.*
5. *Let A be a nonnegative number consisting of integral part* $[A]$ *and fractional part* ν, *i.e.,* $A = [A] + \nu$, *with* $0 \le \nu < 1$. *If m is a positive integer, we may write*

$$[mA] = m[A] + [m\nu], \quad with \ 0 \le [m\nu] < m.$$

Hence the following relationship, which will prove valuable upon discussing base-conversion algorithms:

$$[mA] - m[A] = ([mA] \bmod m).$$

PASCAL'S DIVISIBILITY TEST

Pascal attempted to devise an algorithm intended to determine if arbitrary integer γ is divisible by some divisor d, also an integer, which algorithm would be simpler than long division, and make use of the properties of uniform base m positional number systems.

If symbol δ_i^γ denotes the ith digit of γ, base m, integer γ is expressed as

$$\gamma = (\delta_{n-1}^\gamma, \ldots, \delta_i^\gamma, \ldots, \delta_2^\gamma, \delta_1^\gamma, \delta_0^\gamma \cdot)_m$$

Pascal's method proceeds in two steps:

1. We first turn our attention to *divisor* d: the residues modulo d of the successive powers of m are extracted in sequence. Let these be

$$\rho_0 = (m^0 \bmod d) \equiv 1, \quad \rho_1 = (m^1 \bmod d),$$
$$\rho_2 = (m^2 \bmod d), \ldots, \rho_{n-1} = (m^{n-1} \bmod d).$$

   We shall call the infinite sequence $\mathbf{R}(d)_m = (\rho_0, \rho_1, \rho_2 \ldots)$ the *residue sequence for divisor* d, *base* m. (Regardless of base m under consideration, integer d between brackets will always be written in decimal notation.) We shall return to the residue sequence in more detail, following a discussion of Euler's theorem.

2. Having generated the initial n terms of the residue sequence for divisor d, we now turn to *dividend* γ. Clearly, γ is divisible by d if and only if

$$P \equiv \sum_{i=0}^{n-1} \delta_i^\gamma m^i \equiv \sum_{i=0}^{n-1} \delta_i^\gamma \rho_i = 0 \; (\bmod d).$$

   By analogy to the scalar product of two vectors, we shall call P the *Pascal product* of the residue sequence $\mathbf{R}(d)_m$ and the base m representation of γ.

In order to determine whether or not integer γ of length n is divisible by integer d, the representation of γ, base m is written facing the initial n terms of the residue sequence $\mathbf{R}(d)_m$. That may conveniently be done from right to left, consistent with Arabic notation. The top panel of Figure 3.1 shows how the divisibility of 396 by 44 is tested in base 2. Applying Pascal's algorithm, we multiply the numbers facing each other and add modulo 44. We get $P \equiv 88 \equiv 0 \pmod{44}$, indicating that 44 divides 396. As a further example, the procedure was repeated with $m = 5$, for $\gamma = 6646$ and $d = 44$ (bottom panel

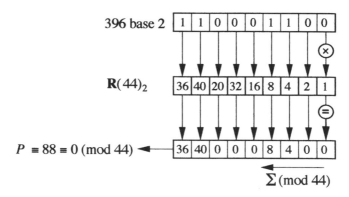

Divisibility of 396 by 44 (base 2)

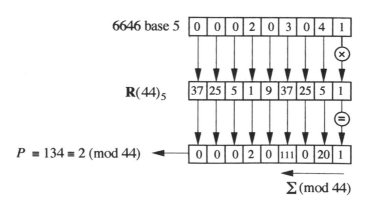

Divisibility of 6646 by 44 (base 5)

Figure 3.1. Pascal's algorithm for testing divisibility. Divisibility of 396 by 44, base 2. Divisibility of 6646 by 44, base 5.

of Figure 3.1). This gives $P \equiv 134 \equiv 2 \pmod{44}$, signifying that the division of 6646 by 44 leaves residue 2.

A particularly useful case is that of divisor 9, whose base 10 residue sequence is $\mathbf{R}(9)_{10} = (1, 1, 1, 1, \ldots)$. An integer's modulo 9 residue is therefore congruent to the modulo 9 sum of its base 10 integers (and the sum of the integers of that sum, and so on). That statement is the foundation of the method of "*casting out nines*," which was so popular until the emergence of electronic hand-held calculators. That method was imported to Europe by the Arabs, along with the Hindu numerals, and was described by al-Khawarizmy and others in their reckoning manuals. It was shown that the arithmetic operations of addition, subtraction, and multiplication could be verified by repeating the same operation with the operand's residues, and then checking that the residues of both operations' results were identical. For example, to the difference $(1994 - 1789)$ corresponds the modulo 9 difference $(1 + 9 + 9 + 4) - (1 + 7 + 8 + 9) \equiv 5 - 7 \equiv -2 \equiv -2 + 9 = 7 \pmod 9$, which is congruent to $205 \pmod 9$. Similarly, to the product (1994×1789) corresponds $5 \times 7 = 35 = 8 \pmod 9$, which is congruent to $3\,567\,266 \pmod 9$.

Checking of arithmetic operations may be performed modulo any arbitrary integer, and in his *Liber abaci*, Leonardo Fibonacci advocated using integers 11 or 7 in addition to 9. The merit of modulus 11 obviously stems from the fact that the residue sequence $\mathbf{R}(11)_{10}$ is $(1, -1, 1, -1, \ldots)$. Extracting the residue modulo 11 of a base 10 number is thus obtained by forming the sum of its even-rank digits and that of its odd-rank digits, and then subtracting the second sum from the first, and repeating that operation on the resulting integer again and again, until a residue between 0 and 10 is obtained. For example, $1994 \equiv (4 + 9) - (9 + 1) \equiv 3 \pmod{11}$, and $1789 \equiv (9 + 7) - (8 + 1) \equiv 7 \pmod{11}$, yielding $7 \times 3 \equiv 21 \equiv 10 \pmod{11}$. On the other hand, we also have $1789 \times 1994 = 3\,567\,266 \equiv (6 + 2 + 6 + 3) - (6 + 7 + 5) \equiv 10 \pmod{11}$.

For numbers written in the binary system, the residue sequence $\mathbf{R}(3)_2$ is also $(1, -1, 1, -1, \ldots)$. Base 2 operations may thus be easily checked modulo 3, using the same principle as above. For example $367 = (1\,0\,1\,1\,0\,1\,1\,1\,1.)_2 \equiv 4 - 3 \equiv 1 \pmod 3$.

One can imagine feeding the "bit stream" corresponding to a given number into a special kind of binary counter containing only two positions, as in Figure 3.2. That counter is incremented by one bit with every even-rank 1-bit input, and by two bits with every odd-

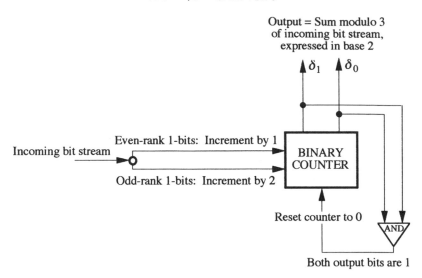

Figure 3.2. Modulo 3 checking device for a binary bit stream.

rank 1-bit input. It instantly resets itself to 0 whenever both positions contain ones. Incrementing by two bits is equivalent to decrementing by one bit, modulo 3. Resetting to 0 is the materialization of $3 \equiv 0$ (mod 3). At the end of the operation, the counter displays the input stream's mod 3 sum.

As a manual method aimed at ascertaining whether or not an integer divides another, the Pascal algorithm may at first glance appear simpler than long division, but it quickly reaches a point of diminishing returns. Whether or not it is indeed simpler remains in the eye of the beholder, unless one develops some objective criteria for the measurement of algorithm complexity that apply to the above. Its study allows us, nonetheless, to gain useful insight into the properties of divisibility, which properties will be called upon later in this chapter.

EULER'S FUNCTION AND THEOREM

Euler's function $\phi(d)$ is defined as the number of integers, from 1 to $d - 1$, that are relatively prime to the integer d. These integers are referred to as the *relatively prime residues* of d. For example, the relatively prime residues of 7 are 1, 2, 3, 4, 5, 6, yielding $\phi(7) = 6$, and those of 10 are 1, 3, 7, 9, yielding $\phi(10) = 4$.

Leonhard Euler. Portrait by Emanuel Handmann. Courtesy Oeffentliche
Kunstsammlung Basel, Kunstmuseum. Photo: Martin Bühler.

Euler was the first to offer a general solution, around 1760, to
the problem of determining the number of relatively prime residues
of any given integer. He showed that for a composite number A, for
example, whose different prime factors are a, b, c, regardless of their
multiplicities,

$$\phi(A) = A\left(1 - \frac{1}{a}\right)\left(1 - \frac{1}{b}\right)\left(1 - \frac{1}{c}\right). \qquad (3.3)$$

To understand how he achieved that result, let us partition the A integers (0 to $A-1$) into a class containing the multiples of a, which are
A/a in number, and a class containing the remaining integers, which
are $A\left(1 - \frac{1}{a}\right)$ in number, and then discard the first class. Let us now
partition the remaining class into a class containing the $A\left(1 - \frac{1}{a}\right)\left(\frac{1}{b}\right)$
multiples of b and a class containing the $A\left(1 - \frac{1}{a}\right)\left(1 - \frac{1}{b}\right)$ remaining integers, and then discard the first class. Finally, let us partition
the remaining class into a class containing the multiples of c and a
class not containing those multiples, and then discard the former,
leaving us with the class of integers that are relatively prime to A,
whose number is $A\left(1 - \frac{1}{a}\right)\left(1 - \frac{1}{b}\right)\left(1 - \frac{1}{c}\right)$. For example, $\phi(10) =$
$10(1 - 1/2)(1 - 1/5) = 4$, $\quad \phi(44) = 44(1 - 1/11)(1 - 1/2) = 20$.

For any prime p, we get $\phi(p^n) = p^n\left(1 - \dfrac{1}{p}\right)$, and consequently $\phi(p) = p - 1$. Clearly, $\phi(1)$ has no meaning, since the integer 1 has no prime factors. However, by special definition,[2] one puts $\phi(1) = 1$.

Euler's Theorem

Consider the congruence

$$m^y \equiv 1 \pmod{d} \qquad m > 1, y > 0. \tag{3.4a}$$

If m and d are not relatively prime, it may be established, as follows, that there is no nonzero exponent y that satisfies congruence (3.4a). If there existed some value a of y for which $m^a \equiv 1 \pmod{d}$, we would have $m^a \equiv 1 \pmod{d_0}$ and $m^a \equiv 1 \pmod{\bar{d}}$ (where, as defined earlier, $d = d_0 \bar{d}$ and d_0 is the largest factor of d that is relatively prime to m). The second congruence implies that $m^y \equiv 1 \pmod{\bar{d}}$ for every y multiple of a. That is not possible, since, by definition, $m^y \equiv 0 \pmod{\bar{d}}$ for all values of y equal to or larger than the exponent of m over d.

On the other hand, if m and d are relatively prime, we may legitimately ask ourselves if there always exists some nonzero value of y that satisfies congruence (3.4a). The answer was provided by Euler's theorem, which states that for any integer m other than 1 that is relatively prime to d,

$$m^{\phi(d)} \equiv 1 \pmod{d}. \tag{3.4b}$$

Instead of providing a general proof of congruence (3.4b), we shall illustrate the proof's construction by means of an example.

Proof—Specific case. Let $d = 8$ and $m = 15$. The relatively prime residues of 8 are 1, 3, 5, 7, signifying that $\phi(8) = 4$. We now calculate the modulo 8 residues of integer 15 multiplied by each of the four prime residues of 8, and call the products *subresidues*. We obtain

$$(15 \times 1) \equiv 7 \pmod{8}, \qquad (15 \times 3) \equiv 5 \pmod{8},$$

$$(15 \times 5) \equiv 3 \pmod{8}, \qquad (15 \times 7) \equiv 1 \pmod{8}.$$

[2]Oysten Ore, *Number Theory and Its History* (New York: Dover, 1988), p. 110.

We observe the following:

1. Since integer 15 and every residue by which it is multiplied are relatively prime to 8, so is each subresidue.
2. Since the set $\{1, 3, 5, 7\}$ contains every residue mod 8 that is prime to 8, it follows that the subresidues all belong to that set.
3. Suppose that two subresidues are the same, for example, that $15a \equiv 15b \pmod 8$, where a and b are different prime residues of 8. We know that 15 and 8 are relatively prime. Factor 15 may therefore be canceled, resulting in $a \equiv b$, which is contrary to our hypothesis. Therefore no two subresidues may be the same. We are in presence of four different subresidues.
4. Since the number of different subresidues is equal to the number of residues, and all subresidues belong to the set $\{1, 3, 5, 7\}$ of residues, it follows that the set of subresidues is identical to the residue set.
5. Since the congruences we calculated are all modulo 8, we may multiply all their left-hand members as well as their right-hand members. Thus

$$15^4 \times (1 \times 3 \times 5 \times 7) \equiv (1 \times 3 \times 5 \times 7) \pmod 8,$$

and since $(1 \times 3 \times 5 \times 7)$ is relatively prime to 8, it may be canceled from both sides, resulting in

$$15^{\phi(8)} \equiv 1 \pmod 8.$$

Q.E.D.

Euler's theorem is a generalization of Fermat's so-called *little theorem*, which states that for any prime p and any integer a not a multiple of p,

$$a^{p-1} \equiv 1 \pmod p.$$

Putting $a = 2$, it follows that for any prime number other than 2,

$$2^{p-1} \equiv 1 \pmod p.$$

That statement provides a shortcut to determining whether any number is *composite*. Whereas the statement $2^{x-1} \equiv 1 \pmod{x}$ may be verified for nonprime numbers, such as $x = 341$, it remains the case that if $(2^{x-1} \bmod x)$ is different from 1, we know that x is not prime (it is composite). In 1986, Adelman, Rumely, Cohen, and Lenstra developed the eponymous ARCL test, which determines with certainty whether a number of up to fifty digits is prime. The test is predicated on Fermat's theorem, and when it is run on fast number-crunchers, it provides an answer in seconds.

Exponents

An important and easily proved theorem states that the smallest power y of m that satisfies congruence (3.4a) divides $\phi(d)$. If t is the smallest such power, m is said to belong to exponent $t \pmod{d}$, and t is called the *exponent to which m belongs* $(\bmod\, d)$, or simply the *exponent of m* $(\bmod\, d)$. As it turns out in the previous example, integer $4 = \phi(8)$ is not the smallest power t such that $15^t \equiv 1 \pmod 8$. That power is 2, and we conclude that 15 belongs to exponent 2 $(\bmod\, 8)$.

Euler's theorem may thus be reformulated as follows:

For every integer $m > 1$, a positive integer y can be found such that $m^y \equiv 1 \pmod{d}$ if and only if m and d are relatively prime. The smallest such integer divides Euler's number $\phi(d)$.

(Clearly, any multiple y of that smallest integer also satisifies the congruence, in particular, $y = \phi(d)$.)

I am grateful to Professor Donald E. Knuth for bringing to my attention the fact that a remarkable generalization of Euler's theorem (for relatively prime m and d) was given by Gauss in his *Disquisitiones arithmeticae* (secs. 90–92). That generalization was obviously little known, since it is often ascribed to R. D. Carmichael, who rediscovered it in 1910. The interested reader may turn to this chapter's appendix for that generalization.

It can also be easily seen that a number m is relatively prime to d if and only if it is congruent modulo d to a prime residue of d. If t is the exponent $(\bmod\, d)$ of some prime residue of d, it is also the exponent $(\bmod\, d)$ of any number m that is congruent modulo d to that prime residue. For example, $7^2 \equiv 15^2 \equiv 23^2 \equiv \cdots \equiv 1 \pmod 8$.

According to Euler's theorem, every prime residue of d belongs (mod d) to some exponent t that divides $\phi(d)$. Is the converse of that statement true? In other words, is *every* divisor of $\phi(d)$ the exponent (mod d) of some prime residue of d?

i. If the modulus d is not prime, the answer is no, as illustrated by Table 3.1. The modulus $d = 15$, whose Euler number is $\phi(15) = 8$, has relatively prime residues $1, 2, 4, 7, 8, 11, 13, 14$. Of these, residue 1 trivially belongs to exponent 1 (mod 15), residues $4, 11, 14$ belong to exponent 2 (mod 15), residues $2, 7, 8, 13$ belong to exponent 4, and there is no relatively prime residue of 15 that belongs to exponent 8. On the other hand, the relatively prime residues of 10 are 1, 3, 7, 9. Of these, residue 1 trivially belongs (mod 10) to exponent 1, residue 9 belongs to 2, and residues 3, 7 belong to $\phi(10) = 4$. Remember that for every prime residue q of d, the smallest integer y that satisfies $q^y \equiv 1$ (mod d) necessarily divides $\phi(d)$, implying $q^{\phi(d)} \equiv 1$ (mod d). When we say in Table 3.1 that there is no prime residue of 15 that *belongs* to $\phi(d)$, we mean that there is no q such that $\phi(d)$ is the *smallest* exponent y satisfying $q^y \equiv 1$ (mod d). Obviously, any prime residue q of 15 satisfies $q^8 \equiv 1$ (mod 15).

ii. If modulus d is prime, it may be shown that the answer is yes, in other words, that there exist residues belonging to every divisor of its Euler number $\phi(p) = p - 1$, as illustrated by Table 3.2.

TABLE 3.1
Nonprime modulus.

Divisor t of $\phi(10) = 4$	Prime residues of 10 belonging to t (mod 10)	Divisor t of $\phi(15) = 8$	Prime residues of 15 belonging to t (mod 15)
1	1	1	1
2	9	2	4, 11, 14
4	3, 7	4	2, 7, 8, 13
		8	none

TABLE 3.2
Prime modulus.

Divisor t of $\phi(7) = 6$	Residues of 7 belonging to t (mod 7)	Divisor t of $\phi(13) = 12$	Residues of 13 belonging to t (mod 13)
1	1	1	1
2	6	2	12
3	2, 4	3	3, 9
6	3, 5	4	5, 8
		6	4, 10
		12	2, 6, 7, 11

Primitive Roots

It may be shown that if m belongs to exponent t (mod p), then $(m^q \bmod p)$ also belongs to t (mod p) for every relatively prime residue q of t. For example, Table 3.2 shows that the class of relatively prime residues of 7 that belong to exponent 3 (mod 7) consists of residues 2 and 4, with $2^2 \equiv 4$, and $4^2 \equiv 2$ (mod 7), where $q = 2$ is the only relatively prime residue of exponent 3 other than 1. Similarly, the class of relatively prime residues of 13 that belong to exponent 6 (mod 13) consists of integers 4 and 10, with $4^5 \equiv 10$ and $10^5 \equiv 4$ (mod 13), where integer 5 is the only relatively prime residue of exponent 6, other than 1. The class of prime residues of 13 that belong to exponent 4 (mod 13) consists of integers 5 and 8, with $5^3 \equiv 8$ and $8^3 \equiv 5$ (mod 13), where 3 is the only prime residue of exponent 4 other than 1.

An integer which belongs to $\phi(d)$ (mod d) is called a *primitive root* of d. For example, 3, 7 are primitive roots of 10, and 2, 6, 7, 11 are primitive roots of 13, whereas 15 has no primitive roots. It follows that any integer that is congruent (mod d) to a primitive root of d is also a primitive root of d. Any listing of primitive roots need consist only of relatively prime residues. The class of primitive roots of a prime number p has $\phi(p - 1)$ members. If ϕ represents any one such root, then $(\phi^q \bmod p)$ is a *distinct* member of that class for every value assigned to q from within the class of prime residues of $(p - 1)$.

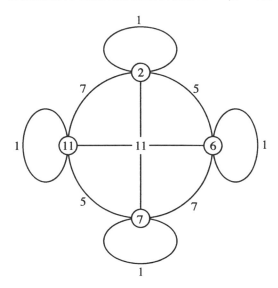

Figure 3.3. Primitive roots of 13 and prime residues of 12.

The primitive roots of 13 are 2, 6, 7, 11. They are four in number, and there are four relatively prime residues of 12, namely, 1, 5, 7, 11. Figure 3.3 illustrates the manner in which each of the four primitive roots of 13 may be obtained from the other residues in its class. The circled numbers in the figure are the primitive roots of 13, and the uncircled numbers are the relatively prime residues of 12. The paths between little circles symbolize the following relationships, mod 13 :

$$2^5 \equiv 6 \quad \text{and} \quad 6^5 \equiv 2 \qquad 6^7 \equiv 7 \quad \text{and} \quad 7^7 \equiv 6$$
$$6^{11} \equiv 11 \quad \text{and} \quad 11^{11} \equiv 6 \qquad 2^{11} \equiv 7 \quad \text{and} \quad 7^{11} \equiv 2$$
. . .

Each path can be followed in both directions, and there exists a path between any two primitive roots. For every prime modulus, it is therefore sufficient to know its *smallest* primitive root, which is usually found by trial and error, as there is no known algorithm for generating it for any given p. That explains why tables of smallest primitive roots are invariably found in every number theory textbook. The *Canon arithmeticus*, published in 1839 by K. G. Jacobi, listed the primitive roots for all primes less than 100. Other, more recent tables go far beyond that. Table 3.3 lists the primitive roots for all primes less than 200.

TABLE 3.3

Smallest primitive roots for
prime modulus $p < 200$

p	SPR	p	SPR
2	1	89	3
3	2	97	5
5	2	101	2
7	3	103	5
11	2	107	2
13	2	109	6
17	3	113	3
19	2	127	3
23	5	131	2
29	2	137	3
31	3	139	2
37	2	149	2
41	6	151	6
43	3	157	5
47	5	163	2
53	2	167	5
59	2	173	2
61	2	179	2
67	2	181	2
71	7	191	19
73	5	193	5
79	3	197	2
83	2	199	3

Note that 10 is a primitive root of 7, since 3 is a primitive root, and $10 \equiv 3 \pmod 7$. Similarly, 10 is a primitive root of 17, since 3 is a primitive root, and $3^3 \equiv 10 \pmod{17}$. That leads us to the question, Which nonprime numbers have primitive roots? We have already observed that integer 10 has two primitive roots among its relatively prime residues, namely, 3, 7. The reader may also verify that integers 2, 5 are primitive roots of 9. Generally, it can be shown that only the following three classes of moduli have primitive roots: (1) $d = p^a$ is the power of an odd prime; (2) $d = 2p^a$; (3) $d = 2$ and $d = 4$.

A Generalization of Euler's Theorem

We now offer an original generalization of Euler's theorem, when m and d are any two integers. Examine the congruence

$$m^{x+y} \equiv m^x \pmod{d} \qquad m > 1, \; x \geq 0, \; y > 0. \qquad (3.5a)$$

According to Euler's theorem, if m and d are relatively prime, $x = 0$ and y is any multiple of the exponent to which m belongs $(\bmod\, d)$, obviously including $\phi(d)$. If m and d are any two positive integers, does congruence (3.5a) have a solution?

Theorem (3.4) states that if $ak = bk \pmod{d}$, the factor k may be canceled from both sides as follows:

$$a \equiv b\left(\bmod\, \frac{d}{(k, d)}\right),$$

where (k, d) is the greatest common divisor of k and d. The factor m^x may thus be canceled from both sides of congruence (3.5a) as follows:

$$m^y \equiv 1\left(\bmod\, \frac{d}{(m^x, d)}\right). \qquad (3.5b)$$

It follows from Euler's theorem that congruence (3.5b), and consequently congruence (3.5a), may have positive solutions in y if and only if $d/(m^x, d)$ is relatively prime to d. The smallest value of x for which that is achieved is the exponent of m over d, in which case $d/(m^x, d) = d_0$, and congruence (3.5b) becomes

$$m^y \equiv 1 \pmod{d_0}. \qquad (3.5c)$$

Since m and d_0 are, by definition, relatively prime, congruence (3.5c) is verified if and only if y is equal to the exponent to which m belongs $(\bmod\, d_0)$, or any multiple thereof. Hence the following theorem:

Given any positive integer pair m, d, the congruence $m^{x+y} \equiv m^x$ (mod d) is verified for positive values of y if and only if $x \geq s$, where s is the exponent of m over d, and y is a multiple of the exponent to which m belongs (mod d_0), where d_0 is the largest factor of d that is relatively prime to m.

It obviously follows that

$$m^{s+\phi(d_0)} \equiv m^s \ (\mathrm{mod}\, d). \qquad (3.5d)$$

If m and d are relatively prime, as stipulated in Euler's theorem, $d_0 = d$ and $x = 0$, and we fall back on Euler's original congruence,

$$m^{\phi(d)} \equiv 1 \ (\mathrm{mod}\, d).$$

It is in that sense that we have taken the liberty of (perhaps presumptuously) referring to (3.5d) as a generalization of Euler's theorem, when it is really a corollary of that important theorem, which extends the theorem's domain to non–relatively prime m and d.

Example 3.2

With $m = 2$ and $d = 36$, we get

$$d = 9 \times 2^2 = 36, \quad d_0 = 9, \quad s = 2,$$

and the exponent of 2 $(\mathrm{mod}\, 9) = 6 = \phi(9)$.
Hence $2^{a+6b} \equiv 2^a \ (\mathrm{mod}\, 36)$ for any integer $a \geq 2$ and any integer b. In particular, $2^{2+6} \equiv 2^2 \ (\mathrm{mod}\, 36)$, $2^{3+12} \equiv 2^3 \equiv 8 \ (\mathrm{mod}\, 36)$, $2^{9+6} \equiv 2^9 \equiv 8 \ (\mathrm{mod}\, 36)$, etc.

The Residue Sequence

As we calculate residue modulo d of the successive powers m^i of m for $i = 0, 1, 2, 3 \ldots$, we necessarily come upon a first value r of i such that the residue m^r is congruent to some earlier residue m^s, since the complete residue set modulo d is finite (by sheer common sense, or by virtue of Dirichlet's distribution principle, discussed in the Marginalia section). In other words, m^s is the first residue that repeats itself. The residue sequence, denoted $\mathbf{R}(d)_m$, thus generally consists of an initial sequence of length s, which we call the *lead sequence*, or *leader*, followed by a *cycle* of length $t = r - s$, which repeats itself indefinitely. If we come upon residue $m^s \ (\mathrm{mod}\, d) \equiv 0$ for some value s of i, all subsequent residues are also zero. That situation occurs when

$d_0 = 1$. In that case, t is indeterminate, and, by convention, we shall put $t = 1$, consistent with $\phi(1) = 1$.

The residue sequence procedure may thus be regarded as an algorithm for generating the smallest values of integers x, y, namely, s, t. In the examples that follow, a colon marks the end of the leader, and a comma marks the end of the first cycle, as in equation (3.6), where ρ_i denotes the modulo d residue of m^i.

$$\mathbf{R}(d)_m = (\rho_0\, \rho_1\, \cdots \rho_{s-1} \colon \rho_s\, \rho_{s+1} \cdots \rho_{s+t-1}, \rho_{s+t} \cdots). \quad (3.6)$$

Example 3.3

    m, d are relatively prime $\bar{d} = 1$, $s = 0$

    $\mathbf{R}(13)_2 = (1\ 2\ 4\ 8\ 3\ 6\ 12\ 11\ 9\ 5\ 10\ 7, 1 \ldots)$

        $t = 12 = \phi(13)$

    $\mathbf{R}(13)_6 = (1\ 6\ 10\ 8\ 9\ 2\ 12\ 7\ 3\ 5\ 4\ 11, 1 \ldots)$

        $t = 12 = \phi(13)$

    $\mathbf{R}(17)_3 = (1\ 3\ 9\ 10\ 13\ 5\ 15\ 11\ 16\ 14\ 8\ 7\ 4\ 12\ 2\ 6, 1\ 3 \ldots)$

        $t = 16 = \phi(17)$

    $\mathbf{R}(7)_4 = (1\ 4\ 2, 1\ 4\ 2 \ldots)$

        $t = 3 = \phi(7)/2$

    $\mathbf{R}(7)_{10} = \mathbf{R}(7)_3 = (1\ 3\ 2\ 6\ 4\ 5, 1\ 3\ 2\ 6 \ldots)$

        $t = 6 = \phi(7)$

Example 3.4

    m, d are not relatively prime, $(\bar{d} \neq 1)$

    $\mathbf{R}(56)_{60} = \mathbf{R}(56)4 = (1\ 4 \colon 16\ 8\ 32, 16\ 8\ 32, \ldots)$

        $d_0 = 7, s = 2, t = 3 = \phi(7)/2$

    $\mathbf{R}(30)_3 = (1 \colon 3\ 9\ 7\ 21, 3 \ldots)$

        $d_0 = 10, s = 1, t = 4 = \phi(10)$

$$\mathbf{R}(36)_2 = (1\ 2 : 4\ 8\ 16\ 32\ 28\ 20, 4\ 8\ \ldots)$$
$$d_0 = 9, s = 2, t = 6 = \phi(9)$$

$$\mathbf{R}(120)_{60} = (1\ 60 : 0, \ldots)$$
$$d_0 = 1, s = 2, t = 1 = \phi(1)$$

The above examples verify Euler's theorem and its generalization.

Indices

The top left panel of Figure 3.4 shows the residues of sequence $\mathbf{R}(13)_2$ strung clockwise around a circle, like the beads of a necklace. Starting from some residue, say, 3, and traveling clockwise around the circle, we say that the *distance* (mod 13, base 2) of residue 6 from residue 3 is 1, that of residue 12 is 2, that of residue 10 is 6, and so on.

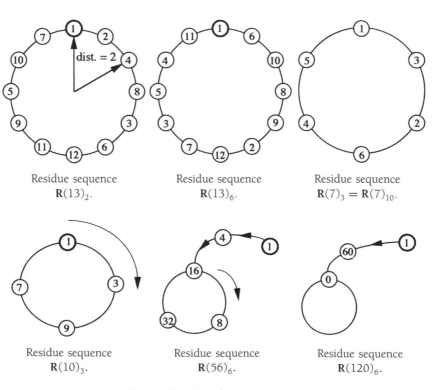

Residue sequence   Residue sequence   Residue sequence
$\mathbf{R}(13)_2$. $\mathbf{R}(13)_6$. $\mathbf{R}(7)_3 = \mathbf{R}(7)_{10}$.

Residue sequence   Residue sequence   Residue sequence
$\mathbf{R}(10)_3$. $\mathbf{R}(56)_6$. $\mathbf{R}(120)_6$.

Figure 3.4. Residue sequences.

Similarly, the distance (mod 13, base 6) of residue 6 from residue 3 is 5, that of residue 12 is 10, and so on.

The distance (mod $p - 1$, base m) from residue 1 of any given residue r of prime number p is called the *index* of r (mod d, base m) and is written $\text{Ind}_m(r)$ (mod d).

It follows that $\text{Ind}_m(1) = 0$ and $\text{Ind}_m(m) = 1$ for any m and d, and we may write

$$r \equiv m^{\text{Ind}_m(r)} \pmod{d}, \tag{3.7}$$

which bears a striking resemblance to the definition of base m logarithms,

$$a = m^{\log_m(a)}. \tag{3.8}$$

Similarly, to the identity $\log(ab) = \log a + \log b$ corresponds the congruence

$$\text{Ind}_m(ab) \equiv \text{Ind}_m a + \text{Ind}_m b \pmod{d} \tag{3.9}$$

If it is understood that m is a number's *smallest primitive root*, we may drop the suffix m and write

$$\text{Ind}(8) = 3 \qquad \text{Ind}(3) = 4 \qquad \text{Ind}(12) = 6 \qquad (\text{mod } 13),$$

where the smallest primitive root of 13 is 2.

Though Euler discovered indices, it was Gauss who offered the first exhaustive analysis in his *Diquisitiones*. Several tables of indices were published over the years. Tables 3.4, 3.5, and 3.6 show the indices for $p = 7$, 13, and 23, and their respective smallest primitive roots, namely, 3, 2, and 5.

TABLE 3.4
Indices for $p = 7$.

Number	1	2	3	4	5	6
Index	0	2	1	4	5	3

TABLE 3.5
Indices for $p = 13$.

Number	1	2	3	4	5	6	7	8	9	10	11	12
Index	0	1	4	2	9	5	11	3	8	10	7	6

TABLE 3.6
Indices for $p = 23$.

Number	1	2	3	4	5	6	7	8	9	10	11
Index	0	2	16	4	1	18	19	6	10	3	9

Number	12	13	14	15	16	17	18	19	20	21	22
Index	20	14	21	17	8	7	12	15	5	13	11

Observe that $\text{Ind}(p - 1) = (p - 1)/2$. That property will be called upon in the discussion of cyclic numbers. From the above discussion, it is clear that the residue sequence $\mathbf{R}(13)_2$ can be rewritten as

$$\mathbf{R}(13)_2 = (1\ 2\ 4\ 8\ 3\ 6\ -1\ -2\ -4\ -8\ -3\ -6, 1\ldots),$$

suggesting that the sum of any two residues whose distance $(\text{mod}(d - 1))$ is equal to $(d - 1)/2$ is zero, and confirming that the sum of the residues modulo d for a primitive root base is divisible by d.

CONJUGATES AND CONFORMABLE MULTIPLES

Congruence (3.5) signifies that the number $m^x(m^y - 1)$ is divisible by d if and only if $x \geq s$ and y is a multiple of t, as defined in the generalization of Euler's theorem. Within the framework of a base m positional number system, any such number takes the form

$$m^x(m^y - 1) = \left(\underbrace{m - 1, \ldots, m - 1}_{y \text{ times } (m-1)}, \underbrace{0, \ldots, 0}_{x \text{ zeros}} \right)_m, \qquad (3.10)$$

that is, y times digit $(m - 1)$ preceded by x zeros, as we move left of the dot. Such a number will be referred to as a *conformable multiple of d*. (The number is written in Arabic order, that is, from right to left, with the radix dot at its right.)

Putting $x = s$ and $y = t$, we shall write

$$m^s(m^t - 1) = M(d)_m, \tag{3.11a}$$

and refer to that number as the *minimum conformable multiple of d, base m*.

We now show that $m^s(m^t - 1)$ divides $m^S(m^T - 1)$ if and only if $S \geq s$ and T is a multiple of t. Indeed, $m^s(m^t - 1)$ divides $m^S(m^T - 1)$ if and only if m^s divides m^S and $(m^t - 1)$ divides $(m^T - 1)$. The first requirement is met if and only if $S \geq s$.

If T is a multiple of t, let $T = at$. This yields

$$\frac{(m^{at} - 1)}{(m^t - 1)} = \sum_{i=0}^{a-1} (m^t)^i, \tag{3.11b}$$

which is an integer, meaning that $(m^t - 1)$ divides $(m^T - 1)$.

If T is not a multiple of t, let $T = at + \rho$, where $\rho < t$, and a is a nonnegative integer. This eventually yields

$$\frac{(m^{at+\rho} - 1)}{(m^t - 1)} = \left(m^\rho \sum_{i=0}^{a-1} (m^t)^i \right) + \left(\frac{m^\rho - 1}{m^t - 1} \right).$$

The first member of the right-hand side is an integer, but the second is not because $\rho < t$, from which it follows that $(m^t - 1)$ does not divide $(m^T - 1)$. Therefore, the second requirement, namely, that $m^t - 1$ divide $m^T - 1$ can be met if and only if T is a multiple of t.

Clearly, then, the minimum conformable multiple of any given d divides any conformable multiple of d. The ratio between two such numbers takes on a striking form, which is invariant with respect to the chosen base.

Example 3.5

$$\frac{2^{20} - 1}{2^5 - 1} = (1\ 0\ 0\ 0\ 0\ 1\ 0\ 0\ 0\ 0\ 1\ 0\ 0\ 0\ 0\ 1.)_2,$$

$$\frac{10^{20} - 1}{10^5 - 1} = (1\ 0\ 0\ 0\ 0\ 1\ 0\ 0\ 0\ 0\ 1\ 0\ 0\ 0\ 0\ 1.)_{10},$$

$$\frac{3^{12} - 1}{3^4 - 1} = (1\ 0\ 0\ 0\ 1\ 0\ 0\ 0\ 1.)_3,$$

and in general, it follows from (3.11b) that

$$\frac{(m^{at} - 1)}{(m^t - 1)} = (1\ 0 \ldots 0\ 1\ 0 \ldots 0\ 1\ 0 \ldots 0\ 1.)_m.$$

The number between parentheses consists of a string of a ones, with consecutive ones separated by $t - 1$ zeros. If, additionally, $S > s$, a number of zeros equal to $S - s$ is added before the base dot, as in

$$\frac{3^5(3^{12} - 1)}{3^3(3^4 - 1)} - (1\ 0\ 0\ 0\ 1\ 0\ 0\ 0\ 1\ 0\ 0.)_3.$$

We shall put

$$\frac{M(d)_m}{d} = \frac{m^s(m^t - 1)}{d} = (d^*)_m \qquad (3.12a)$$

and call integer $(d^*)_m$, or d_m^*, or simply d^* when base m is clear from the context, the *conjugate* of d, base m. The following examples correspond to the residue sequences listed above.

Example 3.6

$$M(13)_2 = (111\ 111\ 111\ 111.)_2 = 2^{12} - 1 = 4\,095$$
$$(13^*)_2 = 4095/13 = 315 = (000\ 100\ 111\ 011.)_2$$

$$M(17)_3 = (2\ 222\ 222\ 222\ 222\ 222.)_3 = 3^{16} - 1 = 43\,046\,720$$
$$(17^*)_3 = 43\,046\,720/17 = 2\,532\,160$$
$$= (0\ 011\ 202\ 122\ 110\ 201.)_3$$

$$M(7)_4 = (3\ 3\ 3.)_4 = 4^3 - 1 = 63$$
$$(7^*)_4 = 63/7 = 9 = (2\ 1.)_4$$

$$M(7)_{10} = 999\,999 = 10^6 - 1$$
$$(7^*)_{10} = 999\,999/7 = 142\,857$$

$$M(56)_{60} = (59\,59\,59\,00.)_{60} = 60^2(60^3 - 1) = 777\,596\,400$$
$$(56^*)_{60} = 777\,596\,400/56 = 13\,885\,650 = (1\ 4\ 17\ 7\ 30.)_{60}$$

$$M(30)_3 = (22\,220.)_3 = 3(3^4 - 1) = 240$$
$$(30^*)_3 = 240/30 = 8 = (22.)_3$$

$$M(36)_2 = (11\ 111\ 100.)_2 = 2^2(2^6 - 1) = 252$$
$$(36^*)_2 = 252/36 = 7 = (111.)_2$$

$$M(120)_{60} = (59\ 0\ 0.)_{60} = 60^2(60 - 1) = 212\,400$$
$$(120^*)_{60} = 212\,400/120 = 1\,770 = (29\ 30.)_{60}$$

As an exercise, express in base 10 the smallest mutiple of 58 consisting of a string of nines followed by a string of zeros.

Ans. $m = 10$, $d = 58$, $\bar{d} = 2$, $d_0 = 29$, $s = 1$, $t =$ the exponent of 10 (mod 29) $= 28$ (10 is a primitive root of 29).

$$58 = \frac{99\ 999\ 999\ 999\ 999\ 999\ 999\ 999\ 999\ 990.}{1\ 724\ 137\ 931\ 034\ 482\ 758\ 620\ 689\ 655.}.$$

The above fraction's numerator is $M(58)_{10}$, and its denominator is $(58^*)_{10}$.

POSITIONAL REPRESENTATION OF RATIONAL NUMBERS

Let us now turn to the inverse $\dfrac{1}{d}$ of integer d. From (3.12a), we have

$$\frac{1}{d} = \frac{d^*}{m^s(m^t - 1)}. \tag{3.12b}$$

Consider the number

$$D = \frac{m^s}{d} = \frac{d^*}{(m^s - 1)}, \tag{3.13a}$$

and assume that it consists of argument $[D]$ and fractional part ν, that is, assume $D = [D] + \nu$. If we multiply D by m^t, we get

$$Dm^t = \frac{d^* m^t}{(m^t - 1)} = d^* m^t \left(\frac{1}{m^t} + \frac{1}{m^{2t}} + \frac{1}{m^{3t}} + \cdots \right)$$

$$= d^* \left(1 + \frac{1}{m^t} + \frac{1}{m^{2t}} + \frac{1}{m^{3t}} + \cdots \right) = d^* + \frac{d^*}{(m^t - 1)} \qquad (3.13b)$$

$$= D + d^* = [D] + d^* + \nu.$$

Since d^* is an integer, the argument of Dm^t is $[D] + d^*$, and its fractional part is ν. That fractional part therefore remains unchanged no matter how many times D is multiplied by m^t. If we write the number D in a base m system, its mantissa thus remains unchanged every time we move the radix dot by t positions to the right. Similarly, $D = m^s/d$ was obtained from $1/d$ by moving the radix dot once by s positions to the right in the latter's representation. We conclude that the base m representation of $1/d$ is periodic, with period (or cycle length) t, and leader length s.

It is worth observing that in any base m number system, regardless of the value of m, we have

$$\frac{1}{m^s(m^t - 1)}$$

$$= \left(\underbrace{\cdot 00 \ldots 0}_{s \text{ zeros}}, \underbrace{00 \ldots 01}_{t-1 \text{ zeros, 1 one}}, \underbrace{00 \ldots 01}_{t-1 \text{ zeros, 1 one}}, \cdots \right)_m, \qquad (3.14a)$$

and it follows from (3.12b) that

$$\frac{1}{d} = d^* \times \left(\underbrace{\cdot 00 \ldots 0}_{s \text{ zeros}}, \underbrace{00 \ldots 01}_{t-1 \text{ zeros, 1 one}}, \underbrace{00 \ldots 01}_{t-1 \text{ zeros, 1 one}}, \cdots \right)_m. \qquad (3.14b)$$

The digit sequence between the parentheses may therefore be referred to as the *pattern*, or *template*, of $1/d$.

Example 3.7

$$d = 56, \text{ base } m = 60 : s = 2, t = 3,$$

$$M(56)_{60} = 60^2(60^3 - 1) = 777\,596\,400.$$

$$(56^*)_{60} = 777\,596\,400/56 = 13\,885\,650 = (1\ 4\ 17\ 7\ 30.)_{60}$$

$$\frac{1}{56} = (1\ 4\ 17\ 7\ 30.)_{60} \times (.0\ 0\ 0\ 0\ 1\ 0\ 0\ 1\ 0\ 0\ 1 \ldots)_{60}$$

$$= (.1\ 4\ 17\ 7\ 30)_{60} + (.0\ 0\ 0\ 1\ 4\ 17\ 7\ 30)_{60} \cdots$$

$$+ (.0\ 0\ 0\ 0\ 0\ 1\ 4\ 17\ 7\ 30)_{60} + \cdots$$

$$= (.1\ 4\ 17\ 8\ 34\ 17\ 8\ 34\ 17\ 8\ 34 \ldots)_{60} = (.1\ 4\ \underline{17\ 8\ 34})_{60}$$

Obviously, this base 60 expression of 1/56 could have been derived in a more straightforward fashion. The purpose of the example was to illustrate the role of the residue (mod d) sequence in predicting the number's *structure*.

Consider now the rational number a/d.

If we write

$$\frac{a}{d} = \frac{(ad^*)}{m^s(m^t - 1)},$$

and put

$$D' = aD = \frac{am^s}{d} = \frac{(ad^*)}{(m^t - 1)}, \tag{3.15}$$

steps (3.13a) and (3.13b) may be exactly repeated, replacing d^* by (ad^*), and D by D'. The period t of rational number a/d is therefore the same as that of $1/d$, and only the lead length s may be different. It follows that the base m mantissa of fraction a/d is such that its lead sequence length s is the exponent of m over d, and its period t is the exponent to which m belongs (mod d_0). When m and d are relatively prime, $s = 0$ and $d_0 = d$.

We have already shown that periodic representations in any uniform positional system correspond to rational numbers. We have just proven the converse of that statement, namely, that the representation of a rational number is always periodic, and its period depends solely on the denominator of whatever fraction (reduced or not) represents that rational number. We have also devised a practical algorithm for the determination of its leader and cycle lengths, which was justified on theoretical grounds.

Mixed Bases

Pascal's divisibility test was elaborated within the framework of uniform periodic bases, decimal or otherwise. Nothing in the above analysis precludes extending the test to any base, mixed or uniform, periodic or not. The residue sequence is obtained as follows:

$$\mathbf{R}(d)_b = (\rho_0 \, \rho_1 \, \rho_1, \ldots),$$

with $\rho_0 = 1$, $\quad \rho_1 = (\pi_1 \bmod d)$, $\quad \rho_2 = (\pi_2 \bmod d), \ldots$

The n-digit integer γ is divisible by d if and only if

$$P \equiv \sum_{i=0}^{n-1} \delta_i^\gamma \pi_i \equiv \sum_{i=0}^{n-1} \delta_i^\gamma \rho_i \equiv 0 \; (\bmod \, d). \tag{3.16}$$

For example, base

$$b = (\ldots 7, 3, 7, 3, 7, 3.),$$
$$\gamma = 58608 = (2\ 0\ 2\ 6\ 0\ 6\ 0)_b, \quad d = 44;$$

this gives $P \equiv 176 \equiv 0 \; (\bmod \, 44)$, indicating that 44 divides 58608 (Figure 3.5).

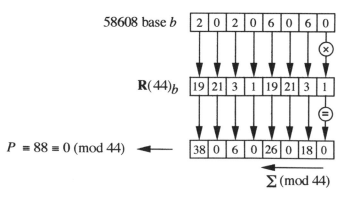

Figure 3.5. Divisibility of 58608 by 44, base $b = (\ldots 3, 7, 3, 7, 3.)$.

As we calculate the residues modulo d of π_i for consecutive values of i, we necessarily come upon a first value r of i such that the residue of π_i is congruent (mod d) to some earlier residue of π_s, where $(r - s)$ is a multiple of base period τ. Putting $t = a\tau = r - s$, we get

$$\pi_{s+a\tau} \equiv \pi_s \ (\mathrm{mod}\, d). \tag{3.17}$$

We can now construct the number $\pi_{s+a\tau}$, base b, and write it facing the residue sequence $\mathbf{R}(d)_b$. (Both sequences are written backwards, in Arabic order.)

$$\pi_{s+a\tau}: \ \ldots \ 0 \ \ (m_{s+a\tau-1} - 1) \ldots (m_{s+1} - 1)\,(m_s - 1) \ \ 0 \ \ \ldots \ \ 0 \ \ 0 \ \ 0.$$

$$\downarrow \qquad \downarrow \qquad\qquad \downarrow \qquad \downarrow \qquad \downarrow \qquad\quad \downarrow \ \ \downarrow \ \ \downarrow$$

$$\mathbf{R}(d)_b: \ \ldots (\pi_s) \ \ (\pi_{s+a\tau-1}) \quad \ldots \quad (\pi_{s+1}) \quad (\pi_s) \ \ (\pi_{s-1}) \ldots (\pi_2)\,(\pi_1)\,(\pi_0)$$

The Pascal product of the above two sequences is

$$P \equiv \pi_s(m_s - 1) + \pi_{s+1}(m_{s+1} - 1) + \cdots$$
$$+ \pi_{s+a\tau-1}(m_{s+a\tau-1} - 1) \ (\mathrm{mod}\, d).$$

If we write the positive terms of the above product facing the corresponding negative terms, we obtain

$$P \equiv \quad \pi_{s+1} + \pi_{s+2} + \cdots + \pi_{s+a\tau-1} + \pi_{s+a\tau}$$
$$\qquad\qquad\qquad\qquad\qquad\qquad\qquad\qquad (\mathrm{mod}\, d).$$
$$- \pi_s - \pi_{s+1} - \pi_{s+2} - \cdots - \pi_{s+a\tau-1}$$

Given that $\pi_{s+a\tau} \equiv \pi_s \ (\mathrm{mod}\, d)$, we get $P \equiv 0 \ (\mathrm{mod}\, d)$; in other words, d divides $\pi_{s+a\tau}$.

As in the case of uniform bases, the residue sequence algorithm allows the determination of the minimum value s of y and the minimum value a of x that satisfy the congruence

$$\pi_{x+\tau y} \equiv \pi_x \ (\mathrm{mod}\, d). \tag{3.18}$$

Example 3.8

i. Base $b = (\ldots 6, 5, 2, 6, 5, 2 .)$: $\quad \tau = 3$, $\pi_\tau = 60$

$\mathbf{R}(7)_b = (1\ 2\ 3\ 4\ 1\ 5\ 2\ 4\ 6, 1\ 2\ 3\ 4 \ldots)$
$\quad s = 0, t = 9 = 3\tau,$
$M(7)_b = (5\ 4\ 1\ 5\ 4\ 1\ 5\ 4\ 1 .)_b = 215\,999$
$\quad d^* = 30\,857.$

$\mathbf{R}(14)_b = (1 : 2\ 10\ 4\ 8\ 12\ 2\ 4\ 6\ 8, 2 \ldots)$
$\quad s = 1, t = 9 = 3\tau,$
$M(14)_b = (1\ 5\ 4\ 1\ 5\ 4\ 1\ 5\ 4\ 0 .)_b = 431\,998.$
$\quad d^* = 30\,857.$

$\mathbf{R}(21)_b = (1\ 2\ 10 : 18\ 15\ 12\ 9\ 18\ 6\ 15\ 9\ 3, 18 \ldots)$
$\quad s = 3, t = 9 = 3,$
$M(21)_b = (5\ 4\ 1\ 5\ 4\ 1\ 5\ 4\ 1\ 0\ 0\ 0 .)_b = 12\,959\,940.$
$\quad d^* = 617\,140.$

$\mathbf{R}(56)_b = (1\ 2\ 10\ 4 : 8\ 40\ 16\ 32\ 48\ 8\ 16\ 24\ 32, 8 \ldots)$
$\quad s = 4, t = 9 = 3\tau,$
$M(56)_b = (1\ 5\ 4\ 1\ 5\ 4\ 1\ 5\ 4\ 0\ 0\ 0\ 0 .)_b = 25\,919\,880.$
$\quad d^* = 462\,855.$

ii. Base $c = (\ldots 5, 4, 3, 5, 4, 3 .)$ $\tau = 3$, $\pi_\tau = 60$

$\mathbf{R}(11)_c = (1\ 3\ 1\ 5\ 4\ 5\ 3\ 9\ 3\ 4\ 1\ 4\ 9\ 5\ 9, 1 \ldots)$
$\quad s = 0, t = 15 = 5\tau, M(11)_c = 777\,599\,999$
$\quad d^* = 70\,690\,909.$

$\mathbf{R}(220)_c = (1\ 3\ 12 : 60\ 180\ 60\ 80\ 20\ 80\ 180\ 100\ 180\ 20$
$\qquad\qquad\qquad\qquad\qquad 60\ 20\ 100\ 80\ 100, 60 \ldots)$
$\quad s = 3, t = 15 = 5\tau, M(220)_c = 46\,655\,999\,940.$
$\quad d^* = 212\,072\,727.$

$\mathbf{R}(90)_c = (1\ 3\ 12\ 60 : 0\ 0\ 0, 0\ 0\ 0 \ldots)$
$\quad s = 4, t = 3 = \tau, M(90)_c = 777\,599\,999.$
$\quad d^* = 70\,690\,909.$

We may now return to Euler's theorem and our generalization thereof, within the context of periodic mixed-base systems. If π_τ and d are relatively prime, $s = 0$, and a is none other than the exponent to which π_τ belongs (mod d). In particular, remembering that $(\pi_\tau)^y = \pi_{\tau y}$, we get

$$\pi_{\tau\phi(d)} \equiv 1 \ (\mathrm{mod}\, d). \tag{3.19}$$

If π_τ and d are not necessarily relatively prime, the generalization of Euler's theorem becomes in this case:

The congruence $\pi_{x+y} \equiv \pi_x \pmod d$ is verified for nonzero values of y if and only if $x \geq s$, and y is a multiple of $a\tau$, where s is the smallest integer such that π_s is divisible by d/d_0, and a is the exponent to which π_τ belongs $\pmod{d_0}$, where d_0 is the largest factor of d that is relatively prime to π_τ.

In particular, we get

$$\pi_{s+\tau\phi(d_0)} \equiv \pi_s \pmod d. \qquad (3.20)$$

Clearly, $\pi_{s+a\tau}$ is a conformable multiple of d. But is it the minimum such multiple? The above two examples were deliberately chosen because the answer to that question is not necessarily affirmative, as we may encounter a residue's reoccurrence at a distance not multiple of τ. Indeed, $\rho_4 = \rho_0 = 1$ in $\mathbf{R}(7)_b$, and $\rho_2 = \rho_0 = 1$ in $\mathbf{R}(11)_c$. It follows that the conformable four-digit base b integer $(1\ 5\ 4\ 1\ .)_b = \pi_4 - 1 = 119$ is divisible by 7. Similarly, the conformable two-digit base c integer $(3\ 2\ .)_c = \pi_2 - 1 = 11$ is obviously divisible by 11. We also observe that $\rho_1 = \rho_6 = 2$ in sequence $\mathbf{R}(7)_b$, from which it follows that the number $(5\ 4\ 1\ 5\ 4\ 0\ .)_b = 3598$ is also a multiple of 7.

What we are looking for, therefore, are true minimum conformable multiples of d, whose length $a\tau$ is a multiple of τ. With that in mind, we shall say that the true minimum conformable multiple of d is $M(d)_m = \pi_{s+a\tau}$, and its base b conjugate is

$$d^* = \frac{M(d)_m}{d} = \frac{\pi_{s+a\tau}}{d}, \qquad (3.21)$$

from which it follows that the period of the reciprocal of d is $a\tau$, and its lead length is s.

In conclusion, within a periodic base, mixed or uniform, a rational number's representation is always periodic, and conversely, any periodic representation always corresponds to a rational number.

Bases 2 and 10

Tables 3.7 and 3.8 give the numbers s and t for bases 2 and 10, with d ranging from 1 to 50.

TABLE 3.7

Leader length s and period t, base 2.

d	d_0	$\phi(d_0)$	s	t	d	d_0	$\phi(d_0)$	s	t
1	1	1	0	1	26	13	12	1	12
2	1	1	1	1	27	27	18	0	18
3	3	2	0	2	28	7	6	2	3
4	1	1	2	1	29	29	28	0	28
5	5	4	0	4	30	15	8	1	4
6	3	2	1	2	31	31	30	0	5
7	7	6	0	3	32	1	1	5	1
8	1	1	3	1	33	33	20	0	10
9	9	6	0	6	34	17	16	1	8
10	5	4	1	4	35	35	24	0	12
11	11	10	0	10	36	9	6	2	6
12	3	2	2	2	37	37	36	0	36
13	13	12	0	12	38	19	18	1	18
14	7	6	1	3	39	39	24	0	12
15	15	8	0	4	40	5	4	3	4
16	1	1	4	1	41	41	40	0	20
17	17	16	0	8	42	21	12	1	6
18	9	6	1	6	43	43	42	0	14
19	19	18	0	18	44	11	10	2	10
20	5	4	2	4	45	45	24	0	12
21	21	12	0	6	46	23	22	1	11
22	11	10	1	10	47	47	46	0	23
23	23	22	0	11	48	3	2	4	2
24	3	2	3	2	49	49	42	0	21
25	25	20	0	20	50	25	20	1	20

Example 3.9

i. Base 2.
$$d = 20 : \quad s = 2, t = 4, d^* = 3 = (1\ 1.)_2$$
$$1/d = (1\ 1.)_2 \times (.0\ 0\ \underline{0\ 0\ 0\ 1})_2 = (.0\ 0\ \underline{0\ 0\ 1\ 1})_2$$
$$d = 36 : \quad s = 2, t = 6, d^* = (1\ 1\ 1.)_2$$
$$1/d = (1\ 1\ 1.)_2 \times (.0\ 0\ \underline{0\ 0\ 0\ 0\ 0\ 1}.)_2$$
$$= (.0\ 0\ \underline{0\ 0\ 0\ 1\ 1\ 1}.)_2$$

TABLE 3.8
Leader length s and period t, base 10.

d	d_0	$\phi(d_0)$	s	t	d	d_0	$\phi(d_0)$	s	t
1	1	1	0	1	26	13	12	1	6
2	1	1	1	1	27	27	18	0	3
3	3	2	0	1	28	7	6	2	6
4	1	1	2	1	29	29	28	0	28
5	1	1	1	1	30	3	2	1	1
6	3	2	1	1	31	31	30	0	15
7	7	6	0	6	32	1	1	5	1
8	1	1	3	1	33	33	20	0	2
9	9	6	0	1	34	17	16	1	16
10	1	1	1	1	35	7	6	1	6
11	11	10	0	2	36	9	6	2	1
12	3	2	2	1	37	37	36	0	3
13	13	12	0	6	38	19	18	1	18
14	7	6	1	6	39	39	24	0	6
15	3	2	1	1	40	1	1	3	1
16	1	1	4	1	41	41	40	0	5
17	17	16	0	16	42	21	12	1	6
18	9	6	1	1	43	43	42	0	21
19	19	18	0	18	44	11	10	2	2
20	1	1	2	1	45	9	6	1	1
21	21	12	0	6	46	23	22	1	22
22	11	10	1	2	47	47	46	0	46
23	23	22	0	22	48	3	2	4	1
24	3	2	3	1	49	49	42	0	42
25	1	1	2	1	50	1	1	2	1

ii. Base 10.

$d=20$: $s=2, t=1, d^*=45$

$$1/d = 45 \times .0\ 0\ \underline{1} = .045 + .0045 + .00045 + \cdots$$
$$= .04\underline{9} = .5$$

$d=36$: $s=2, t=1, d^*=25$

$$1/d = 25 \times .0\ 0\ \underline{1} = .0\ 2\ 5 + .0\ 0\ 2\ 5 + .0\ 0\ 0\ 2\ 5 + \cdots$$
$$= .0\ 2\ \underline{7}$$

A correct expression of $1/d$ may still be obtained if instead of taking the minimum conformable multiple and the conjugate of d, we take any conformable multiple M of d, together with the corresponding *pseudo-conjugate* M/d. In the fourth example above, $s = 2$, $t = 1$, $d^* = 25$. Instead, we may take $s = 3$, $t = 2$, which yields pseudo-conjugate $99000/36 = 2750$, and we get

$$1/d = 2750 \times .0\ 0\ 0\ \underline{01}$$
$$= .0\ 2\ 7\ 5 + .0\ 0\ 0\ 2\ 7\ 5 + .0\ 0\ 0\ 0\ 0\ 2\ 7\ 5 + \cdots = .0\ 2\ \underline{7}.$$

Cyclic Numbers

An interesting diversion is offered by cyclic numbers, in light of Euler's theorem. A cyclic number is an n-digit base m number that, when multiplied by any integer from 1 to n, results in a number whose digits are those of the original number in the same cyclic order, albeit shifted around the loop formed by joining the number's ends.

Figure 3.6a lists the twelve configurations resulting from the base 2 multiplication of integer 315 by multipliers 1 through 12; the corresponding multiplier ρ_i is listed on the left of the configuration. Figure 3.7a shows these configurations not in order of increasing multipliers, but in such a manner that each configuration is shifted by one position to the left with respect to its predecessor. Figure 3.6b lists the configurations corresponding to $(12\ 37\ 50\ 25\,.\,)_{63}$. The most

$\rho_0 = 1$	0 0 0 1 0 0 1 1 1 0 1 1	
$\rho_1 = 2$	0 0 1 0 0 1 1 1 0 1 1 0	
$\rho_2 = 4$	0 1 0 0 1 1 1 0 1 1 0 0	
$\rho_3 = 8$	1 0 0 1 1 1 0 1 1 0 0 0	
$\rho_4 = 3$	0 0 1 1 1 0 1 1 0 0 0 1	
$\rho_5 = 6$	0 1 1 1 0 1 1 0 0 0 1 0	
$\rho_6 = 12$	1 1 1 0 1 1 0 0 0 1 0 0	
$\rho_7 = 11$	1 1 0 1 1 0 0 0 1 0 0 1	
$\rho_8 = 9$	1 0 1 1 0 0 0 1 0 0 1 1	
$\rho_9 = 5$	0 1 1 0 0 0 1 0 0 1 1 1	
$\rho_{10} = 10$	1 1 0 0 0 1 0 0 1 1 1 0	
$\rho_{11} = 7$	1 0 0 0 1 0 0 1 1 1 0 1	

Figure 3.6a. $(13^*)_2 \times \rho_i$

$\rho_1 = 3$	37 50 25 12
$\rho_2 = 4$	50 25 12 37
$\rho_3 = 2$	25 12 37 50

Figure 3.6b. $(5^*)_{63} \times \rho_i$

$\rho_0 = 1$	1 4 2 8 5 7
$\rho_1 = 3$	4 2 8 5 7 1
$\rho_2 = 2$	2 8 5 7 1 4
$\rho_3 = 6$	8 5 7 1 4 2
$\rho_4 = 4$	5 7 1 4 2 8
$\rho_5 = 5$	7 1 4 2 8 5

Figure 3.6c. $(7^*)_{10} \times \rho_i$

Figure 3.6. Cyclic numbers.

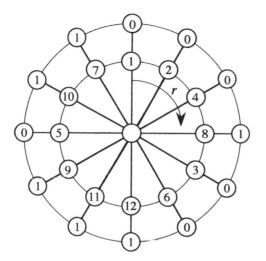

Figure 3.7a. The cyclic number $(13^*)_2$.

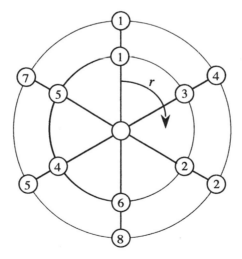

Figure 3.7b. The cyclic number $(7^*)_{10}$.

classical example concerning cyclic numbers is that of base 10 integer 142 857, shown in Figures 3.6c and 3.7b.

The above observations may be explained by analyzing the third example as follows:

$$\frac{1}{7} = \frac{(7^*)_{10}}{999\,999} = \frac{142847}{999\,999} = .142\,857\,142\,857\ldots.$$

Hence

$$\frac{10}{7} = 1 + \frac{3}{7} = 1.428\,571\,428\ldots,$$

and

$$\frac{3}{7} = .428\,571\,428\ldots = \frac{428\,571}{999\,999};$$

i.e.,

$$3 \times \frac{142\,857}{999\,999} = \frac{428\,471}{999\,999},$$

or

$$3 \times 142\,857 = 428\,571.$$

Similarly,

$$2 \times 142\,857 = 285\,714,$$

and so on. In other words, multiplying $(7^*)_{10}$ by residue ρ_i of 7 shifts the cyclic number 142857 by a number of positions equal to index i.

Clearly then, a base m number is cyclic if it is the conjugate p^* of a prime number p, where m is a primitive root of p or is congruent (mod p) to a primitive root of p. Indeed, it is only in that case that the range of values of index i, namely, 0 to $p-2$, is equal to the number of prime residues of p, namely, $p-1$.

Multiplying the cyclic number $(p^*)_m$ by any residue ρ_i of p shifts its digits by i positions. Index i is usually written $\text{Ind}_m(\rho_i)$ (mod p).

It follows that multiplying $(p^*)_m$ by $(p-1)$ shifts it by $\text{Ind}_m(p-1) = (p-1)/2$, and we get

$$(p^*)_m + (p-1)(p^*)_m = p(p^*)_m = M(p)_m = m^p - 1.$$

In other words, shifting $(p^*)_m$ by half its length and adding it to $(p^*)_m$ results in a base m number consisting of a sequence of $(p-1)$ digits, all of which are $(m-1)$.

Example 3.10

    i. $(7^*)_{10} + 6(7^*)_{10} = 7(7^*)_{10}$

          $= 142\,857 + 857\,142 = 999\,999.$

    ii. $(5^*)_{63} + 4(5^*)_{63}$

          $= 5(5^*)_{63} = (12\ 37\ 50\ 25.)_{63} + (50\ 25\ 12\ 37.)_{63}$

          $= (62\ 62\ 62\ 62.)_{63}.$

   iii. $(13^*)_2 + 12(13^*)_2 = 13(13^*)_2$

          $= (111\ 011\ 000\ 100.)_2 + (000\ 100\ 111\ 011.)_2$

          $= (111\ 111\ 111\ 111.)_2.$

The property just described turns out to be a particular case of a theorem dicovered in 1836 by French mathematician Midy, which states that if cycle length t of the decimal mantissa of irreducible fraction a/p (p is prime) is even, the sum of its two halves is a string of nines. That is true even if the mantissas do not correspond to cyclic numbers.

Decimal cyclic numbers occur if and only if integer 10 is a primitive root of some prime p. The reader may verify that those primes smaller than 100 that have integer 10 among their primitive roots, are 7, 17, 19, 23, 29, 41, 47, 59, 61, and 97. There exists no known algorithm for generating these particular primes, no more than there exists a known algorithm for generating primes in general!

According to Martin Gardner, to whom I shall always be indebted for teaching me the joy of mathematics, William Shanks discovered that 17389 is one such prime, and correctly calculated the 17,388 decimal digits of its corresponding cyclic number, though he apparently offered a flawed calculation of π to 707 decimals![3] In *A History of π*, Petr Beckman tells the story of that mishap, along with many other fascinating stories of stubborn calculators through the ages.[4]

Strings of Ones and Zeros

In his book *More Mathematical Puzzles and Diversions*,[5] Martin Gardner poses the problem of determining the smallest number whose decimal representation consists of an uninterrupted string of ones followed by an uninterrupted string of zeros, which number is evenly divisible by a given number d. For example, for $d = 225$, the required number is 11 111 111 100. Generalizing to any uniform base m, the problem consists of finding the smallest integers x, y such that $m^{x+y} - m^x$ is divisible by d and $m - 1$, in other words, of finding coefficient λ such that $m^{s+\lambda t} - m^s$ is divisible by $m - 1$, i.e.,

$$\frac{m^{s+\lambda t} - m^s}{m - 1} \equiv 0 \pmod{d}.$$

Disregarding the leader and adding the integers within one full cycle of length t of residue sequence $\mathbf{R}(d)_m$ results in

$$S = \sum_{i=s}^{s+t-1} m^i = \frac{m^{s+t} - m^s}{m - 1}.$$

Clearly, $m^{s+t} - m^s \equiv 0 \pmod{d}$, and we may be in presence of one of the following cases:

a. If $(m - 1)$ is relatively prime to d, $S \equiv 0 \pmod{d}$. The required number is $M(d)_m/(m - 1)$.
b. If $(m - 1)$ is not relatively prime to d, S may or may not be congruent to zero $\pmod{d}$. If S is congruent to 0, the required number is $M(d)_m/(m - 1)$. Otherwise, it is $(m^{s+\lambda t} - m^s)/(m - 1)$, where λ denotes the smallest integer such that λS is a multiple of d.

---

[3]Martin Gardner, *Mathematical Circus* (New York: Knopf, 1979).
[4]Petr Beckman, *A History of π* (New York: St. Martin's Press, 1971).
[5]Martin Gardner, *More Mathematical Puzzles and Diversions* (Harmondsworth, England: Pelican, 1969), p. 35.

Plate 1. Surveyors measuring the grain crop. Tomb of Menna in West Thebes. Courtesy The American University in Cairo Press.

Plate 2. Fragment of the Rhind Papyrus. Courtesy British Museum, London.

Plate 3. Jean-François Champollion. Portrait by Leon Cogniet (1794–1880), Louvre Museum. © Réunion des Musées Nationaux. Photo R. J. Ojeda. Paris.

Plate 4. Inscription on the temple of Luxor annotated by Champollion. The inscription is an accounting of the war loot of Ramses II, following a fierce battle. Facing each hieroglyph, Champollion wrote the Greek, French, and demotic translations, in descending order. The English translation of the excerpt is, "Wine, measures, six thousand, four hundred, twenty eight. Goats, four thousand, six hundred, twenty two . . ." Reproduced from *Description de l'Egypte* by Champollion.

Plate 5. The Eye of Horus and corresponding unit fractions.

Plate 6. The Egyptian Scribe. Louvre Museum.
© Réunion des Musées Nationaux. Photo H. Lewandowski.

Plate 7. Detail of Mayan Codex. Courtesy Dresden Museum.

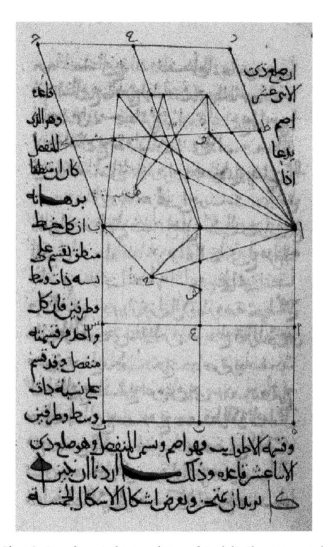

Plate 8. Page from Arabic translation of Euclid's *Elements*, copied
in 1188 by Mas'ud Mohamed ibn Sa'id. Purchased in 1917
at Bukhara by V. A. Ivanov for the Asiatic Museum. Courtesy the
Russian Academy of Sciences and the ARCH Foundation, Lugano.

Plate 9. French revolutionary print showing the new units. From the *Petit décadaire d'instruction publique*, printed during the French Revolution.

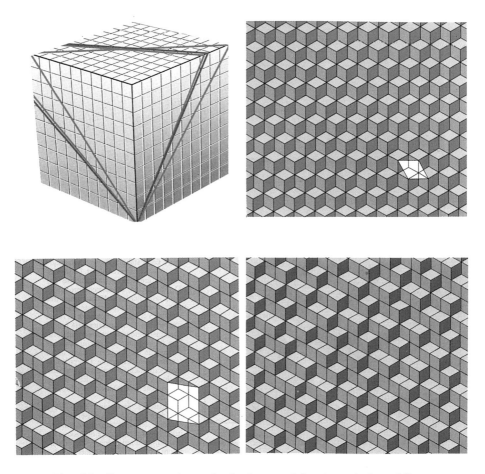

Plate 10. Cleaving a regular stack of cubes. *Top left*, cube with three different cleaving angles. *Top right*, diagonal cut on each of three adjacent faces. *Bottom left*, slope of 4/3 on right-hand face and 2/3 on left-hand face. *Bottom right*, irrational angle on both faces. White parallelograms are capable of regularly tiling the plane. Reproduced from Peter Stephens and Alan Goldman, "La Structure des quasi-cristaux," *Pour la science*, June 1991: 56–63. Editions Belin, Paris.

Example 3.11

$R(7)_{10} = (1\ 3\ 2\ 6\ 4\ 5, 1\ 3\ 2\ 6...): s=0, t=6, S=21, \lambda=1$

The required decimal number is 111 111.

$R(8)_{10} = (1\ 2\ 4:0,0,0...): s=3, t=1, S=0, \lambda=1$

The required decimal number is 1 000.

$R(27)_{10} = (1\ 10\ 19, 1\ 10...): s=3, t=3, S=30, \lambda=9$

The required decimal number is a string of 27 ones.

$R(75)_{10} = (1\ 10:25, 25...): s=2, t=1, S=25, \lambda=3$

The required decimal number is 1 1 1 0 0.

$R(44)_{11} = (1:11\ 33, 11\ 33...): s=1, t=2, S=44, \lambda=1$

The required base 11 number is $(1\ 1\ 0.)_{11} = 132$.

$R(7)_{15} = (1, 1...): s=0, t=1, S=1, \lambda=7$

The required base 15 number is a string of 7 ones,

base 11; i.e., $(15^7 - 1)/14 = 1743463$.

$R(225)_{10} = (1\ 10:100, 100...), : s=2, t=1, S=100, \lambda=9$

The required decimal number is 11 111 111 100.

Finally, the reader may wish to check the following division:

$$58 = \frac{11\ 111\ 111\ 111\ 111\ 111\ 111\ 111\ 111\ 110}{191\ 570\ 881\ 226\ 053\ 639\ 846\ 743\ 295}$$

If that game is played with binary numbers, it is clear that the sought-after number is none other than the divisor's minimum conformable multiple, base 2.

A number consisting of an interrupted string of n ones in a given base is called a *repunit*, and is denoted R_n in that base. Decimal repunits are generally composite. The first prime decimal repunit is R_{19}, and the next is R_{23}.

MARGINALIA

Mersenne Primes

In one of his epistolary exchanges with Bernard Frenicle de Bessy, the French friar Marin Mersenne (1588–1648), also a regular correspondent of Pierre de Fermat, expressed his views on prime numbers of the form $2^n - 1$, within the general context of an ongoing discussion of the so-called perfect numbers (numbers that are equal to the sum of the number's factors, such as $28 = 2^2(2^3 - 1) = 1 + 2 + 4 + 7 + 14$), of which it is known that they are of the form $2^{n-1}(2^n - 1)$, where $2^n - 1$ is prime. The following is a brief discussion of those primes, which are referred to as *Mersenne primes*. Table 3.9 lists the prime factors of $2^n - 1$ for n ranging from 2 to 31.

TABLE 3.9
Prime factors of $2^n - 1$.

n	$2^n - 1$	n	$2^n - 1$
2	3	17	131071
3	7	18	$3^3 \cdot 7 \cdot 19 \cdot 73$
4	$3 \cdot 5$	19	524287
5	31	20	$3 \cdot 5^2 \cdot 11 \cdot 31 \cdot 41$
6	$3^2 \cdot 7$	21	$7^2 \cdot 127 \cdot 337$
7	127	22	$3 \cdot 23 \cdot 89 \cdot 683$
8	$3 \cdot 5 \cdot 17$	23	$47 \cdot 178481$
9	$7 \cdot 73$	24	$3^2 \cdot 5 \cdot 7 \cdot 13 \cdot 17 \cdot 241$
10	$3 \cdot 11 \cdot 31$	25	$31 \cdot 601 \cdot 1801$
11	$23 \cdot 89$	26	$8191 \cdot 8193$
12	$3^2 \cdot 5 \cdot 7 \cdot 13$	27	$7 \cdot 73 \cdot 262657$
13	8191	28	$3 \cdot 5 \cdot 29 \cdot 43 \cdot 113 \cdot 127$
14	$3 \cdot 43 \cdot 127$	29	$233 \cdot 2204167$
15	$7 \cdot 31 \cdot 151$	30	$3^2 \cdot 7 \cdot 11 \cdot 31 \cdot 151 \cdot 331$
16	$3 \cdot 5 \cdot 17 \cdot 257$	31	2147483647

If n is not prime, $2^n - 1$ is divisible by $2^f - 1$, where f is any factor of n. For example,

$$\frac{2^{20} - 1}{2^1 - 1} = (11111111111111111111.)_2 = 3 \cdot 5^2 \cdot 11 \cdot 31 \cdot 41,$$

$$\frac{2^{20} - 1}{2^2 - 1} = (1010101010101010101.)_2 = 5^2 \cdot 11 \cdot 31 \cdot 41,$$

$$\frac{2^{20} - 1}{2^4 - 1} = (10001000100010001.)_2 = 5 \cdot 11 \cdot 31 \cdot 41,$$

$$\frac{2^{20} - 1}{2^5 - 1} = (1000010000100001.)_2 = 3 \cdot 5^2 \cdot 11 \cdot 41,$$

$$\frac{2^{20} - 1}{2^{10} - 1} = (10000000001.)_2 = 5^2 \cdot 41.$$

When n is prime, it has obviously no factor f such that $2^f - 1$ divides $2^n - 1$ (other than n itself). That condition is necessary for $2^n - 1$ to be prime, but it is not sufficient, as $2^n - 1$ may have factors of another form. A simple example is provided by $n = 11$, for which $2^{11} - 1 = 2047 = 23 \times 89$.

Table 3.7 shows for entry $d = 7$ the values $s = 0$, indicating that d is odd, and $t = 3$, meaning that the base 2 minimum conformable multiple of 7 is equal to $2^3 - 1 = 7$, or that 3 is the exponent y that yields the smallest number $2^y - 1$ multiple of 7. That number is none other than 7 itself. Compare that situation to that of $d = 23$. Here, integer 11 is the exponent that yields the smallest number $(2^y - 1)$ multiple of 23. That number is 2047. It is indeed divisible by 23, but it is divisible by 89 as well, whereas $2^3 - 1 = 7$ is divisible only by 7. The number $2^3 - 1 = 7$ is a Mersenne prime. Though 11 itself is a prime number, $2^{11} - 1$ is not a Mersenne prime.

The table reveals the first three Mersenne primes, corresponding to $t = 2, 3, 5, s = 0$, and disqualifies $t = 11, 23$. Table 3.9 reveals the Mersenne primes corresponding to $p = 2, 3, 5, 7, 13, 17, 19, 31$, and disqualifies $p = 11, 23, 29$.

TABLE 3.10

Mersenne primes, or their number of decimal digits, for $p < 100\,000$.
Source: J. H. Conway and R. K. Guy, The Book
of Numbers (New York: Springer Verlag, 1996).

p	$2^p - 1$	p	Digits	p	Digits	p	Digits
2	3	31	(10)	1279	(386)	9941	(2993)
3	7	61	(19)	2203	(664)	11213	(3376)
5	31	89	(27)	2281	(687)	19937	(6002)
7	127	107	(33)	3217	(969)	21701	(6533)
13	8191	127	(39)	4253	(1281)	23209	(6987)
17	131071	521	(157)	4423	(1332)	44497	(13395)
19	524287	607	(183)	9689	(2917)	86243	(25962)

Consider $2^{10} - 1 = 3 \cdot 11 \cdot 31$. That statement signifies that $2^{10} - 1$ is divisible by 3, 11, 31, 33, 93 (in addition, of course, to 1 and 1023). That may be restated as

$$2^{10} \equiv 1 \pmod{3, 11, 31, 33, 93},$$

meaning that the exponent to which 2 belongs modulo any one of these factors must divide 10. Indeed, $2^2 \equiv 1 \pmod 3$, $2^{10} \equiv 1 \pmod{11}$, $2^5 \equiv 1 \pmod{31}$, $2^{10} \equiv 1 \pmod{33}$, $2^{10} \equiv 1 \pmod{93}$.

In general, then, the exponent to which 2 belongs modulo any factor of $2^n - 1$ divides n, and at least one such exponent is n. If n is prime, none of the exponents may obviously divide it, and they are all equal to n. It may therefore be stated that for a prime p, the exponent to which 2 belongs modulo any factor of $2^p - 1$ is p. That statement[6] is equivalent to the classical statement, due to Euler, that any factor of $2^p - 1$ is congruent to 1 modulo p. Both statements may be verified by examining Table 3.9 for every prime value of n.

Table 3.10 shows the first twenty-eight prime exponents corresponding to Mersenne primes, along with the actual value, or the number of decimal digits for each of the latter. (Obviously, the number of binary digits is none other than p.)

[6]The proof is left to the reader as an exercise.

On Dirichlet's Distribution Principle

The following remarks will perhaps give the reader a chance to "take a break" from the implacable logic that is so characteristic of number-theoretic reasoning:

A pool table is equipped with six pockets in the traditional manner. You are given a number of billiard balls, and instructed to drop them one at a time into the pockets, going around the table. If you are given exactly six balls, you end up with exactly one ball in each pocket. If you are given seven balls or more, you eventually reach a point where at least one ball falls into a previously occupied pocket.

Common sense? Perhaps. But not rigorous enough in the eyes of serious mathematicians! In order for the underlying logic to be acceptable in mathematical terms, it has to be formalized, and pronounced in axiomatic form. It probably takes a scientist of established reputation to dare the axiomatization of the obvious, which is what Euclid did when he stated his five postulates and five common notions.

Our pool table axiom was bravely enunciated by the German mathematician Peter Gustav Lejeune Dirichlet, who succeeded Gauss as professor of mathematics at Göttingen University in 1855. In general terms, *Dirichlet's distribution principle* states that if p objects are assigned to q slots, and p is greater than q, then at least one slot will receive more than one object.[7] So far, we have intuitively applied that principle at least twice since the opening of this chapter: when, in connection with the residue sequence, we said, "As we calculate residue modulo d of the successive powers m^i of m ($i = 0, 1, 2, 3 \ldots$), we necessarily come upon a value r of i such that residue m^r is congruent to some earlier residue m^s, as the complete residue system modulo d is finite," and again when, in illustrating the proof of Euler's theorem, we stated, "Since the number of different subresidues is equal to that of residues, and all subresidues belong to the set $\{1, 3, 5, 7\}$ of residues, it follows that the set of subresidues is identical to the set of residues." We will apply that principle again when, upon discussing Euclid's algorithm for the determination of the greatest common divisor of two integers, we shall state, "Residue $a_n = 0$ is bound to eventually occur

---

[7]That simplistic version of the principle was inspired by Tobias Dantzig, *Number, the Language of Science*, 4th ed. (Garden City, N.Y.: Doubleday, 1954).

for some value n, since integer sequence $a_0 > a_1 > a_2 > a_3 \ldots > a_n$ can contain no more than a_0 *positive decreasing* integers."

Goethe beautifully captured that habit of mathematicians, of translating prosaic observations into universal and elegant formal structures, when he said, "Mathematicians are a species of Frenchmen: if you say something to them, they translate it into their own language and presto! it is something entirely different."[8] Rather than "different," I submit that the French themselves would no doubt prefer, "Presto! it is now universal truth!"

APPENDIX

Carmichael's Variation on Euler's Theorem

In what follows, a fascinating discovery due to R. D. Carmichael is presented without proof.[9] But we first need to define the *minimal universal exponent* $\lambda(d)$. Putting

$$d = 2^{\alpha_0} \times p_1^{\alpha_1} \times p_2^{\alpha_2} \times \cdots, \qquad (3.22a)$$

where $2, p_1, p_2, \ldots$ are the prime factors of d, and $\alpha_0, \alpha_1, \alpha_2, \ldots$ are their respective multiplicities, we can define $\lambda(d)$ as follows:

$$\lambda(d) \equiv \mathrm{LCM}\big(\lambda(2^{\alpha_0}), \phi(p_1^{\alpha_1}), \phi(p_2^{\alpha_2}), \ldots\big), \qquad (3.22b)$$

where the acronym LCM denotes the least common multiple, and

$$\lambda(2^1) = 1, \quad \lambda(2^2) = 2,$$
$$\lambda(2^{\alpha_0}) = \frac{1}{2}\phi(2^{\alpha_0}) = 2^{\alpha_0 - 2} \quad \text{for } \alpha \geq 3. \qquad (3.22c)$$

For example, for $d = 2^2 \times 3^3 \times 5^2 \times 7 = 18900$, we get

$$\lambda(d) = \mathrm{LCM}\big(\lambda(2^2), \phi(9), \phi(25), \phi(7)\big) = \mathrm{LCM}(2, 6, 20, 6) = 60,$$

---

[8]John L. Casti, *Complexification* (New York: Harper Collins, 1994).
[9]R. D. Carmichael, *Theory of Numbers* (New York: Wiley, 1914).

and for $d = 561 = 3 \times 11 \times 17$, we get

$$\lambda(561) = \text{LCM}(2, 10, 16) = 80.$$

Now, Gauss and Carmichael discovered that for any relatively prime positive integers m, d,

$$m^{\lambda(d)} \equiv 1 \pmod{d}. \tag{3.23}$$

The minimum universal exponent $\lambda(d)$ is a multiple of the exponent to which m belongs mod d, and may sometimes be considerably smaller than $\phi(d)$. Returning to $d = 561$, we have $\lambda(561) = 80$, whereas $\phi(561) = (2 \times 10 \times 16) = 320$. The reader will observe that $\phi(561)$ is a multiple of $\lambda(561)$. But all of this does not mean that the minimum universal exponent is necessarily the exponent to which any integer m belongs $(\bmod\, d)$. Consider, for example, the congruence $4^{20} \equiv 1 \pmod{561}$. Here, the exponent to which 4 belongs modulo 561 is 20, a divisor of 80.

Modulus 561 has the following remarkable property: since 80 is the minimum universal exponent of 561, any multiple y of 80 satisfies the congruence $m^{y} \equiv 1 \pmod{561}$ for any m that is relatively prime to 561. In particular,

$$m^{560} \equiv 1 \pmod{561}. \tag{3.24}$$

Remembering that $m^{p-1} \equiv 1 \pmod{p}$ for any m that is not a multiple of p, the integer 561 is referred to as a *pseudoprime*, or a *Carmichael number*. The next pseudoprimes are 1105, 1729, 2465, 2821, 6601, 8911, 10585, ... It is conjectured that the number of pseudoprimes is infinite.

CHAPTER 4

Real Numbers

God made the integer, the rest is the work of man.
(Leopold Kronecker)

If you are asked the question "What is a natural number?" you will probably reply, "1, 2, 3, . . . ," and not be aware that your reply was not to the question originally put to you, but to the instruction "Recite the natural numbers." In so doing, you have, unbeknownst to you, revealed your affiliation with the intuitionist school of the ancient Greeks, who believed that numbers, which were the essence of everything, just were! To the ancient Greeks, numbers were god-given, and to recite them was to define them. Upon the substrate of natural numbers, the Greeks built the edifice of rational numbers, which in their eyes had no real existence by themseves, but merely represented proportions, or ratios, between integral quantities. Any two (rational) numbers were said to be commensurate, meaning that a third number, their common measure, could always be found, of which they were exact multiples. The apparently gapless succession of rational numbers was a perfect metaphor for the line in geometry. That situation prevailed until they discovered that *unutterable* numbers also existed, such as the square root of the natural number 2, which were not commensurable with the natural numbers, nor indeed among themselves! That untenable crisis prevailed until Eudoxus of Cnidus, around 360 B.C., offered a definition of irrational numbers predicated upon infinite processes. Whereas a rational number could be defined in terms of two natural numbers, an irrational number was defined through the exhaustion of an infinite number of rational numbers. That very same concept was the foundation of the modern definitions of irrationals given by Weierstrass and Dedekind more than two millenia after Eudoxus. The definition of the *continuum* of real numbers by means of the natural numbers alone, constituted a milestone in mathematical thinking, as it arithmetized mathematical analysis, a discipline dealing with infinitesimals and limiting processes. Kronecker attempted to go even further, and refound mathematics upon *finitary methods*, based upon the rejection of infinite processes and the strict reliance on finite entities.

Again, what is a natural number?

In 1912, in *Intuitionism and Formalism*, L. E. J. Brouwer enunciated the principle of intuitionism, which is founded upon acceptance of an intuitive understanding of natural numbers, akin to that of the ancient Greeks. That position was in sharp contrast to the schools of thought advocated by Gottlob Frege and Bertrand Russell on one side, and by Peano on the other.

Let us return to the Mesopotamian sheepherder. Mathematicians would say that to each sheep corresponded one and only one pebble, and, conversely, to each pebble corresponded one and only one sheep. Inasmuch as one sheep is indiscernible from the next, and that character of indiscernibility also applies to pebbles, the sheepherder matched the *quantity* of pebbles to that of the sheep. Let us assume that that quantity was 100. Before the matching operation began, one may conjecture that the sheepherder knew only of sheep and pebbles, both of which are concrete physical entities. Having performed that matching operation, the sheepherder perhaps became aware of some *emergent* attribute, that of quantity, which was shared by the two populations, namely, 100. As Frege and Russell would put it, the (subordinate) class of sheep and the (subordinate) class of pebbles belong to the same *superordinate class*, the number 100. That approach is based on matching, or cardination.

In 1894, between Frege's publication and its rediscovery by Russell, Peano published his own theory in the *Formulaire de mathématiques*, consisting of five axioms founded on the existence of a first natural number, namely, 1, and that of a successor to every natural number. Peano's *ordinal* approach is based not on *sets*, but on *relations* between objects (transitive asymmetrical relations, to be exact). Its essence is therefore counting, or ordination.

Deciding which of the three approaches is the most satisfactory is a matter of personal choice, and one should not yield to the prevailing paradigm of the moment, which may at times be tyrannically imposed upon the mind.

For centuries, the West had naturally embraced the Pythagorean view, and the mathematical education of our children always began with the arithmetic of the natural numbers. During the second half of the twentieth century, Russell's ideas gradually began gaining ground, and resulted in dramatic revisions of mathematical curricula, where the high priests of New Math established an intellectual hegemony, predicated upon the practice of first instilling in young minds the logic of classes, before teaching the children the natural numbers, which were regarded as almost vulgar corollaries of the noble quantification of classes. It may be a little too early to assess whether the New Math revolution has given birth to a generation of better mathematicians, or even promoted better number competence within the average population, or whether it has turned off large numbers of otherwise promising individuals.

"Later mathematicians will regard set theory as a disease from which one has recovered," said Henri Poincaré.

RATIONAL NUMBERS

The Integral Domain

Imagine a number universe I that contains all the integers, positive as well as negative, along with zero, and excludes other numbers. If any two members of I are added, their sum may be found within the universe itself, which is said to be *closed with respect to addition*. No two members of I are such that their sum lies outside it. Since I contains all the negative integers, it is also closed with respect to subtraction. Similarly, multiplying any two integers yields an integer within I, meaning that the universe is also closed with respect to multiplication.

As a mathematical diversion intended to familiarize the reader with the notion of closure, consider the subset S of I, which contains all integers of the form $(4x + 1)$, and only those, where $x = 0, 1, 2 \ldots$. The integers thus defined are $1, 5, 9, 13, 17, 21, 25, 29, \ldots$. Multiplying any two such integers, say, $(4x' + 1)$ and $(4x' + 1)$, results in the integer $(4y + 1)$, which also belongs to S, where $y = 4xx' + x + x'$. Universe S is therefore closed with respect to multiplication. On the other hand, it is clear that the sum of any two numbers lies outside it. This example illustrates how we can construct "artificial" universes (as opposed to the universe of so-called natural numbers) that are endowed with closure with respect to some operations and not others.

Imagine now that integer a of universe I is divided by integer b. Unless a is a multiple of b (a is trivially a multiple of itself and of integer 1), the number that results from the division is not an integer, meaning that universe I is *not* closed with respect to division.

That situation leads us to ask, What kind of number universe can we construct, using integers or combinations thereof, that is closed with respect to addition, multiplication, *and* division?

Before we proceed, let us dwell for a moment on the properties of universe I, where we seem to take it for granted, on the basis of our daily life experience, that "4 *times* 5 *is equal to* 5 *times* 4," "7 *plus* 3 *is equal to* 3 *plus* 7," and so on. Mathematicians are not comfortable with intuitive notions until they have integrated them within the strict confines of a thorough formal system. Self-evident properties are transformed into axioms, which are anointed as incontrovertible mathematical truths. These in turn serve as foundations for the mathematical edifice that is built upon them, by means of universally accepted logical rules of inference. It was not until the nineteenth century, however, that a set of axioms was proposed for the arithmetic of the universe of the integers. The generally accepted set is the following:

1. *The commutative laws of addition and multiplication.* For all $a, b, \quad a + b = b + a$ and $ab = ba$.
2. *The associative laws of addition and multiplication.* For all $a, b, c, \quad a + (b + c) = (a + b) + c$ and $a(bc) = (ab)c$.
3. *The distributive law.* For all $a, b, c, \quad a(b + c) = ab + ac$.
4. *The additive identity law.* There exists a *zero element* 0 such that for all $a, a + 0 = a$.
5. *The multiplicative identity law.* There exists a *unit element* 1 such that for all $a, a \times 1 = a$.
6. *The additive inverse law.* For every a, there is a number a' such that $a + a' = 0$ (a' is denoted $-a$).
7. *The cancellation law.* For all a, b, k, where $k \neq 0$, $ka = kb$ if and only if $a = b$.

The system thus defined is called the *integral domain*.

The Rational Numbers Field

In the universe Q we are about to construct, number entities will be represented not by integers alone, but by *ordered integer pairs* (a, b), where a and b may be positive or negative, and b is not allowed to be zero. The statement $A = (15, 81)$ will indicate that integer pair

$(15, 81)$ *is a representation* of number A, or, better still, that symbol A and pair $(15, 81)$ are equivalent representations of the same number. Rules for handling integer pairs will be spelled out in terms of the properties of integers—in other words, in terms of the rules that govern the integral domain.

I have arbitrarily chosen to derive the rules governing the manipulation of integer pairs from a set of three elementary additional rules. These may be regarded as *definitions* (or axioms) of equality, multiplication, and addition of number pairs. Whereas any member of I may obviously have only one *representation* in terms of integers, namely, itself, a number belonging to Q may have, as we shall see, an infinite number of *equivalent, though different* integer pair representations.

Rule 1. $(a, b) = (c, d)$ if and only if $ad = bc$,

which is read: Pairs (a, b) and (c, d) are said to be equivalent; in other words, they represent the same number, if and only if $ad = bc$.

It follows from the commutativity of integer multiplication that for any a, b, k, $(ak, bk) = (a, b)$. If a and b are relatively prime, the pair (a, b) is said to be *irreducible*, and any pair (ak, bk) may be reduced to (a, b).

Rule 2. $(a, b) \times (c, d) = (ac, bd)$,

which is read: If (a, b) is a representation of the number A and (c, d) is a representation of the number B, then (ac, bd) is a representation of the number $A \times B$.

Rule 3. $(a, b) + (c, b) = (a + c, b)$.

It follows from rule 1 that for any a, b, c, d, $(a, b) = (ad, bd)$ and $(c, d) = (bc, bd)$. Hence $(a, b) + (c, d) = (ad, bd) + (bc, bd)$, and by virtue of rule 3, we get

$$(a, b) + (c, d) = ((ad + bc), bd).$$

The following body of rules may now be derived using ordinary arithmetic:

1a. *Commutativity of multiplication.* It follows from rule 2 and the commutativity of integer multiplication that $(a, b)(c, d) = (ac, bd) = (ca, db) = (c, d)(a, b)$.

1b. *Commutativity of addition.* $(a, b) + (c, d) = (ad + bc, bd) = (cb + da, bd) = (c, d) + (a + b)$.

2a. *Associativity of multiplication.* From rule 2 and the associative law of integers, $(a, b)\{(c, d)(e, f)\} = (a, b)(ce, df) = (ace, bdf) = (ac, bd)(e, f) = \{(a, b)(c, d)\}(e, f)$.

2b. *Associativity of addition.* $(a, b) + \{(c, d) + (e, f)\} = (adf + bcf + bde,\ bdf) = \{(a, b) + (c, d)\} + (e, f)$.

3. *Distributivity.* $(a, b)\{(c, d) + (e, f)\} = (a, b)\ (c, d) + (a, b)\ (e, f)$. Using a strategy similar to the above, distributivity is easily established.

4a. *Additive identity.* $(a, b) + (0, d) = (ad + b \times 0, bd) = (ad, bd) = (a, b)$. For any d, the pair $(0, d)$ is referred to as the *zero element.*

4b. *Multiplicative identity.* By virtue of rule 1, $(d, d) = (1, 1)$ for any d, and $(a, b)\ (1, 1) = (a, b)$. The pair $(1, 1)$ is the *unit element.*

5a. *Additive inverse.* For any a, b, $(a, b) + (-a, b) = (0, b)$. To any pair (a, b) thus corresponds an additive inverse $(-a, b)$. Again, by virtue of rule 1, $(-a, b) = (a, -b)$, and $(-a, -b) = (a, b)$. The additive inverse of (a, b) is denoted $-(a, b)$.

5b. *Multiplicative inverse.* In this paragraph, $a, b, c, d \neq 0$. From rule 2 it follows that $(a, b)\ (b, a) = (ab, ab) = (1, 1)$. To every number (a, b) thus corresponds the multiplicative inverse (b, a). It follows that for any two pairs (a, b), (c, d), we may write $(a, b)\ \{(c, d)\ (d, c)\} = (a, b)$. Hence $(ad, bc)(c, d) = (a, b)$, and if we write $(ad, bc) = (e, f)$, we get $(e, f)(c, d) = (a, b)$. In other words, for any two pairs (a, b), (c, d), there exists a pair (e, f) such that $(e, f)\ (c, d) = (a, b)$; the pair (e, f) is referred to as the outcome of the *division* of (a, b) by (c, d), which may be written $(a, b)/(c, d) = (e, f)$.

The outcome of the division of any number pair by any other nonzero number pair may thus be found within universe Q, which is seen to be closed with respect to addition (subtraction), multiplication, and division.

We leave it to the reader to establish that in this case, the cancellation law is a corollary of the above laws. It is now clear that the universe I of integers, which does not possess the multiplicative inverse property, is a subset of the universe Q of rational numbers. The reader will have surely recognized that the pair (a, b) is none other than the familiar fraction a/b, where a is the numerator, and b the denominator. The system comprising the rational numbers (including integers), together with the arithmetic operations defined above, is referred to as the rational numbers field. A fraction must be understood to be a mere representation of a rational number, which number may otherwise have several different representations. It is the number, an abstract mathematical being, that is rational, not the fraction, and to operations on rational numbers correspond rules for manipulating fractions, such as those listed above.

The Egyptians utilized a very elaborate system where every fraction was equated to the sum of unit fractions, that is to say, fractions with unit numerator. That system seems to have disappeared with Egyptian civilization. The representation of rational numbers by fractions is generally attributed to Liu Hsin, in China, early in the first century. It was later introduced to the West by the twelfth-century Arab mathematician al-Hassar, and utilized by Ibn Mas'ud al-Kashi (ca. 1430) in his book of arithmetic.

As an exercise, the reader may verify that if p is a prime number, the arithmetic modulo p constitutes a field within the realm of the universe of integers. In particular, the reader will want to establish the multiplicative inverse law, namely, that for all a other than 0, there exists an integer b such that $ab \equiv 1 \pmod{p}$. That law fails when the modulus is not prime.

But have we *really* established the property of closure of the universe Q of rational numbers with respect to multiplication, addition, and division? Unfortunately, things are not so simple! As we shall later see, that property of closure fails when dealing with *infinite* sets of fractions, which sets must be made to also include irrational numbers, in order to restore closure.

A definition of *irrational* numbers by means of rational numbers alone, which is rooted in the notion of infinity, was provided by Dedekind as he stood upon the shoulders of Greek giants, not the least of whom was Eudoxus of Cnidus.

Marginalia: On the Axiomatic Method

The above exercise, which may appear somewhat bewildering at first sight, illustrates in rather simple terms how every mathematical theory starts by laying down an arbitrary, and hopefully consistent, set of axioms, from which a body of nontrivial statements may then be derived, using commonsense logical inference rules. Other combinations of elementary statements could have been imagined, hopefully consistent, and capable of generating the same set of rules. These rules would need to be sufficient to cope with every situation involving number pairs, in other words, *complete*. That *axiomatic* approach was the historical contribution of the early Greeks to mathematics, epitomized by Euclid's magnificent thirteen-volume *Elements*. Around 350 B.C., Euclid enunciated a set of axioms, which for centuries served as the foundations of geometry, until Nikolay Ivanovich Lobachevsky invented in 1829 a different geometry, based on a set of axioms allowing more than one parallel to a line to be drawn from a point outside it. Another *non-Euclidian* geometry was invented in 1854 by Bernhard Rieman, which later allowed Einstein to achieve his monumental contribution to the modern theory of gravitation.

Wrote Ernest Nagel and James R. Newman:

> A climate of opinion was thus generated in which it was tacitly assumed that each sector of mathematical thought can be supplied with a set of axioms sufficient for developing systematically the endless totality of true propositions about the given area of inquiry. Gödel's paper showed this assumption is untenable. He presented mathematicians with the astounding and melancholy conclusion that the axiomatic method has inherent limitations, which rule out the possibility that even the ordinary arithmetic of the integers can ever be fully axiomatized.[1]

In the words of E. C. Titchmarsh, whose beautiful little book entitled *Mathematics for the General Reader* inspired the above commentary, "The method that we have used, that of solving an apparently insoluble problem by re-interpreting it in terms of new numbers, ... is indeed one of the principal sources of progress in mathematical ideas."[2] That method consisted of treating number pairs as if they were a new kind of number, with its own distinctive properties. Another striking instance of the use of number pairs is that of complex numbers, where the pair (a, b) consists of the *real* number a and the *imaginary* number b, which together generate the *complex* number $a + ib$, where i is the square root of minus one.

It seems that students of mathematics have an almost unlimited capacity for integrating new elements within their body of knowledge,

---

[1] Ernest Nagel and James R. Newman, *Gödel's Proof* (New York: New York University Press, 1958).

[2] E. C. Titchmarsh, *Mathematics for the General Reader* (New York: Dover, 1981).

such as the handling of fractions, which gradually becomes a kind of second nature, and upon which they draw without consciously paying attention to the sometimes very complex nature of the subject at hand. Such is also the case of the positional representation of numbers and its accompanying arithmetical algorithms, as well as that of imaginaries, which electronic engineers use daily, probably oblivious of the theoretical foundations upon which they are built.

Commensurability

Despite the relative lack of interest manifested by the early Greeks in material objects surrounding them, they had been thrilled by the discovery of the relation betwen the pitch of a vibrating string and its length, confirming their mystical faith in number and proportionality. Reducing the length of a vibrating string by one-half raises its pitch an octave; by one-third, a fifth; by one-quarter, a fourth; by one-fifth, a major third; by one-sixth, a minor third (Figure 4.1).

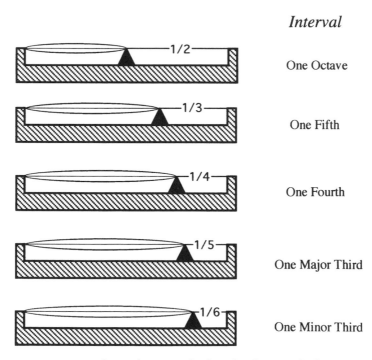

Interval

One Octave

One Fifth

One Fourth

One Major Third

One Minor Third

Figure 4.1. Relation between the length of a stretched string
and musical intervals.

Figure 4.2. A 1492 representation of Pythagoras studying harmony and number. *Source*: F. Gafurius, *Theorica musice* (Milan, 1492).

Said Aristotle in the *Metaphysics*

The so-called Pythagoreans applied themselves to mathematics, and... came to believe that its principles are the principles of everything. And since numbers are by nature first among these principles, they fancied that they could detect in numbers, more than in fire and earth and water, many analogues of what is and what comes to be.... And since they saw further that the properties and ratios of the musical scale are based on numbers, and that the numbers are the ultimate things in the whole physical universe, they assumed the elements of numbers to be the elements of everything, and the whole universe to be a harmony of number.[3]

The ancient Greeks were dedicated to unifying their geometrical and number-theoretic visions of the world, each being a metaphor of the other. To their minds, the universe of rational numbers was a perfect metaphor for the line: If you took a length of rope and cut two lengths a and b at random, a third length c could always be found such that $a = mc$ and $b = nc$, where a and b are rational. Thus a and b were said to be commensurable, their common measure being c. That statement could be rewritten $na = mb$, meaning that

[3]S. Hildebrandt and A. Tromba, *Mathematics and Optimal Form* (New York: Scientific American Library, 1985). By "number," Aristotle meant integers.

equal multiples (equimultiples) of a and b could always be found within their number universe. Again, that statement could be written $a/b = m/n$, signifying that the ratio between any two lengths of rope was always a rational number.

That meant that wherever you effected a cut in a line, you fell upon a rational number. Between any two rational numbers, you could squeeze an infinity of rational numbers. Rational numbers were therefore infinitely dense. Both the line and the universe of rational numbers were deemed *gapless*.

Early Greeks were quite happy with that situation: The number-geometry metaphor was flawless. To every point on a line corresponded a rational number, and there was no geometrical manipulation of the line that did not correspond to some combination of the known arithmetic operations, performed within the universe Q.

That was the prevailing blissful situation until an intellectual earthquake hit the Greek landscape. Some numbers were not rational! (Little did the Greeks suspect that these numbers actually far outnumbered the rational numbers.)

But first, let us pause to examine one of the cornerstones of ancient Greek arithmetic: the theorem of Pythagoras.

IRRATIONAL NUMBERS

Pythagoras's Theorem

> Geometry conceals two great treasures: One is the theorem of
> Pythagoras, the other the division of a line into middle
> and extreme ratio. The first is comparable to a
> measure of gold, and the second to a precious jewel.
> (Kepler [1571–1630])

Pythagoras was born in 569 B.C. in Samos, off the Turkish coast, and, following the advice of his teacher, Thales, traveled to Egypt and Mesopotamia before settling in the southern Italian town of Crotona. There, he established a secret Brotherhood dedicated to the study of what was to be later known as the *quadrivium*, namely, arithmetic, harmony, geometry, and astronomy (then referred to as *astrologia*). With Thales and Pythagoras, the beginnings of Greek science may thus be traced back to the Ionian coast of Asia Minor, and regarded as a natural extension of Egyptian and Babylonian science. Modern-day scholars, having been exposed to recent translations of manuscripts and engravings from both regions, view with suspicion Pythagoras's claim to the theorem that to this day bears his name. It is said that he sacrificed an ox to celebrate his imposture. The story does not say, however, who ate the ox, as his followers were allegedly vegetarians! Pythagoras was otherwise notorious for ascribing his followers' findings to his name, and it is also said that those who revealed brotherhood secrets were murdered by their peers.

The discovery of the *alogon*, that unutterable irrational number, was intolerable to Pythagoreans, who were thence strictly forbidden from revealing its existence to the outside world. Rudy Rucker imagined a movielike scenario, liberally adapted from an allegedly true story, in which the Pythagorean Hippasus, a native of the Italian city of Metapontum, upon revealing the existence of the unnameable, "accidentally" drowns at sea, in the course of an ill-fated brotherhood outing.[4] I shall contribute to that tragic scenario by imagining that a bird of ill omen was flying above the scene of the tragic drowning, and warned Pythagoras that upon fleeing from Croton, he would be murdered—in Metapontum, the very town where poor Hippasus was born.

The famous theorem that Pythagoras ascribed to his name states that the square of the hypotenuse of a right triangle is equal to the sum of the squares of the other two sides. (In Figure 4.3a, the area of the square erected on the hypotenuse is equal to the sum of the areas of the squares erected on the other two sides.) That theorem is proposition I.47 of Euclid's *Elements*.

Rather than reproduce Euclid's proof, we offer an interesting partly algebraic, partly geometric proof, which was discovered by the Indian mathematician Bhaskara (ca. 1114–1185), author of the *Lilavati*, named after his daughter to console her for not marrying, following a flawed prediction by the mathematician of the exact date of her marriage. The proof is predicated on the construction of Figure 4.3b, where

$$z^2 = (y - x)^2 + 4\left(\frac{xy}{2}\right) = x^2 + y^2. \tag{4.1}$$

[4]Rudy Rucker, *Infinity and the Mind* (New York: Bantam, 1983).

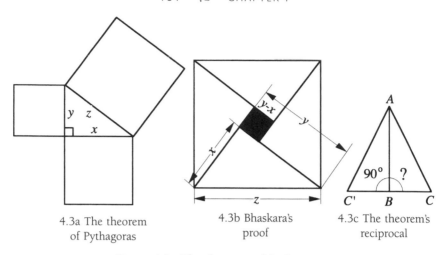

| 4.3a The theorem of Pythagoras | 4.3b Bhaskara's proof | 4.3c The theorem's reciprocal |

Figure 4.3. The theorem of Pythagoras.

While most high school students are capable of reciting the theorem as stated above, few of them are aware that the more exact statement should be that the square of a triangle's side is equal to the sum of the squares of the other two sides *if and only if* it is the hypotenuse of a right triangle. In other words, the reciprocal of the traditional statement is also true. Indeed, proposition I.48, the last proposition of Book I of Euclid's *Elements* states exactly that, and provides a beautiful little proof, illustrated by Figure 4.3c. The proof is based upon Euclid's proposition I.8, which states that if two triangles are such that their three corresponding sides are equal, they are congruent, and consequently their corresponding angles are equal.

Consider Figure 4.3c. By construction, we have $AC^2 = AB^2 + BD^2$. We need to show that ABC is a right angle. To do so, we draw the line $BC' = BC$ perpendicular to AB. Since ABC' is a right triangle, it follows from Pythagoras's theorem that $AC'^2 = AB^2 + BC'^2 = AB^2 + BC^2$, that is, that $AC' = AC$. Triangles ABC and ABC' share side AB. Sides BC and BC' are equal by construction, and we have just shown that $AC = AC'$. By virtue of Euclid's proposition I.18, the triangles are therefore congruent, and ABC is a right angle. Q.E.D.

Pythagorean Triples

According to some historians, ancient Egyptians had discovered that a triangle whose sides measure 3, 4, 5 units contains a right angle, and it is said, albeit not documented, that geometers always carried with them a rope loop upon which twelve knots delimited equal intervals, which allowed them to construct right angles whenever required, as in Figure 4.4.

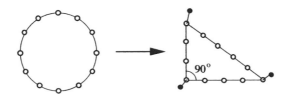

Figure 4.4. The (allegedly) Egyptian knotted rope.

Other examples of integer triples x, y, z, such as $(5, 12, 13)$ and $(9, 40, 41)$ also satisfy the equation $z^2 = x^2 + y^2$. As a matter of fact, there is an infinity of such triples, which are referred to as *Pythagorean*. The triangle they form is also said to be Pythagorean.

It has been established that Indian mathematicians were familiar with Pythagorean triples as early as the second and possibly as early as the fifth century B.C. The text of the *Çulba Sûtra* makes explicit reference to the triples $(3, 4, 5)$, $(5, 12, 13)$, $(7, 24, 25)$, $(8, 15, 17)$, $(12, 35, 37)$, which were used for the construction of right triangles.

Generally, any integer pair a, b $(a > b)$, which may be referred to as *source integers*, generates the Pythagorean triple x, y, z according to the following rule:

$$x = a^2 - b^2, \quad y = 2ab, \quad z = a^2 + b^2. \tag{4.2}$$

For example, integers $a = 3$ and $b = 2$ generate the triple $(5, 12, 13)$.

The reader who is familiar with complex numbers will not have failed to observe that if $v = a + ib$, where $i = \sqrt{-1}$, then $v^2 = x + iy$. Indeed, $v^2 = (a + ib)^2 = a^2 + (ib)^2 + 2iab = (a^2 - b^2) + i(2ab) = x + iy$, and if $\tan \theta = \dfrac{b}{a}$, then

$$\frac{y}{x} = \frac{2 \tan \theta}{1 - \tan^2 \theta} = \tan 2\theta. \tag{4.3}$$

Dividing both sides of equation $z^2 = x^2 + y^2$ by z^2, we get

$$\frac{x^2}{z^2} + \frac{y^2}{z^2} = 1. \tag{4.4}$$

The problem of finding Pythagorean triples is now reduced to that of finding rational number pairs the sum of whose squares is equal to unity.

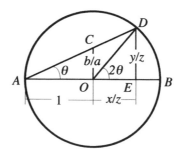

Figure 4.5. Pythagorean triples.

Equation (4.3) suggests the construction shown in Figure 4.5. A circle of radius 1 is drawn with center O. A segment $OC = b/a$ is raised perpendicularly to AB at O. The line AC intersects the circle at D, and we have

$$\frac{DE}{1 + OE} = \frac{b}{a},$$

and

$$DE^2 + OE^2 = 1.$$

The reader may now verify that $DE = y/z$ and $OE = x/z$, where $z^2 = a^2 + b^2$.

It is interesting to point out that Pierre de Fermat, to whom we shall soon return, proved that the area of a Pythagorean triangle, namely $xy/2$, cannot be a perfect square. To achieve that result, he used the method of *infinite descent*, which consisted of showing that if there existed a Pythagorean triple x, y, z such that $xy/2$ was a perfect square, an algorithm would exist that allowed the calculation of a new triple x', y', z', such that $x'y'/2$ was also a perfect square, and $z > z'$. An infinity of ever-smaller integers $z > z' > z'' \ldots$ would therefore also exist, which is obviously impossible.

The Plimpton 322 Tablet

Table 4.1a shows a transcription of the Plimpton Tablet; x and y are the sides of a right triangle, and z is its hypotenuse. The tablet lists, in sexagesimal notation, the values of x and z, along with the ratio z^2/y^2. The tablet was found in relatively good condition, albeit with a few missing pieces, in particular its entire left edge, which most probably contained the digit 1. Its overall consistency authorized its decipherers to fully transcribe it into modern terms. With the exception of lines 2, 9, 13, and 15, the tablet is correct. Erroneous entries are shown in parentheses in Table 4.1a, alongside their corrected values. Table 4.1b shows the decimal translation of a few lines, expliciting the value of sides x, y, and z and ratio z^2/y^2. Side y, the denominator of z/y, has no prime factors other than 2, 3, and 5, thus authorizing finite sexagesimal fractional representations. Line 11

TABLE 4.1a

The Plimpton Tablet, with faults in parentheses.

Line	z^2/y^2	x	z
1	(1), 59, 0, 15	1, 59	2, 49
2	(1), 56, 56, 58, 14, 50, 6, 15	56, 7	(3, 12, 1) 1, 20, 25
3	(1), 55, 7, 41, 15, 33, 45	1, 16, 41	1, 50, 49
4	(1), 53, 10, 0, 29, 32, 52, 16	3, 31, 49	5, 9, 1
5	(1), 48, 54, 1, 40	1, 5	1, 37
6	(1), 47, 6, 41, 40	5, 19	8, 1
7	(1), 43, 11, 56, 28, 26, 40	38, 11	59, 1
8	(1), 41, 33, 59, 3, 45	13, 19	20, 49
9	(1), 38, 33, 36, 36	(9, 1) 8, 1	12, 49
10	(1), 35, 10, 2, 28, 27, 34, 26, 40	1, 22, 41	2, 16, 1
11	(1), 33, 45	45	1, 15
12	(1), 29, 21, 54, 2, 15	27, 59	48, 49
13	(1), 27, 0, 3, 45	(7, 12, 1) 2, 41	4, 49
14	(1), 25, 48, 51, 35, 6, 40	29, 31	53, 49
15	(1), 23, 13, 46, 40	56(53)	1, 46

TABLE 4.1b
Transcription of a line sample from the Plimpton Tablet.

Line	x	y	z	z^2/y^2
1	119	120	169	$428\,415/60^3$
7	2291	2700	3541	$80\,247\,558\,400/60^6$
9	481	600	769	$21\,288\,996/60^4$
11	45	60	75	$5\,625/60^2$
13	161	240	289	$18\,792\,225/60^4$
15	56	90	106	$17\,977\,600/60^4$

TABLE 4.1c
Calculating source integers a, b.

Line	x	y	z	a	b
1	119	120	169	12	5
2	3367	3456	4825	64	27
3	4601	4800	6649	75	32
4	12709	13500	18541	125	54
5	65	72	97	9	4
6	319	360	481	20	9
7	2291	2700	3541	54	25
8	799	960	1249	32	15
9	481	600	769	25	12
10	4961	6480	8161	81	40
11	45	60	75	2	1
12	1679	2400	2929	48	25
13	161	240	289	15	8
14	1771	2700	3229	50	27
15	56	90	106	9	5

corresponds to the famous 3-4-5 Pythagorean triple attributed (probably erroneously) to the Egyptians. It is the first known triple such that $y = x + 1$. Line 1 corresponds to the second such triple, namely, (119, 120, 169). These triples will be later examined in more detail. Table 4.1c shows source integers a, b for every line of the tablet.

The Ladder of Theodorus of Cyrene and
Diophantine Equations

The theorem of Pythagoras led the Greeks of the time to erroneously conjecture that every right triangle had three commensurate, or rational, sides. To their minds, every right triangle was similar to one whose sides were a Pythagorean triple. When that triple was not obvious, they were convinced that an exhaustive search would eventually lead to its discovery. Consider, for example, the isosceles right triangle of side x. The square of its hypotenuse z, according to the irrefutable theorem of Pythagoras, has to be equal to $2x^2$. Imagine now that a laborious Greek mathematics student endeavours to test successive values of the integer x in the hope of eventually discovering some integral value of hypotenuse z that verifies Pythagoras's theorem, in other words, such that $z^2 = 2x^2$. He constructs a table for $x = 1, 2, 3, \ldots$. For each entry, he tests successive values of z, and retains only those values of x yielding the smallest shortfall $|2x^2 - z^2|$. In so doing, he comes upon a succession of *critical* values of x for which the shortfall is narrowed to ± 1, as shown in Table 4.2a. (The skeptical reader may verify that any value of x falling between two critical values of x yields a shortfall $|2x^2 - z^2|$ larger than 1.)

TABLE 4.2a
The Ladder of Theodorus of Cyrene.

n	x	z	$(2x^2 - z^2)$		z/x	
1	1	1	1		1/1	
2	2	3		-1		3/2
3	5	7	1		7/5	
4	12	17		-1		17/12
5	29	41	1		41/29	
6	70	99		-1		99/70
7	169	239	1		239/169	
8	408	577		-1		577/408
9	985	1 393	1		1 393/985	
10	2 378	3 363		-1		3 363/2 378
11	5 741	8 119	1		8 119/5 741	
...						

To make the search easier, his teacher, the Pythagorean Theodorus of Cyrene (Tunisia), reveals to him that the sequence of critical values of x and z may be obtained in a straightforward manner, without fastidiously scanning through every individual value of integer x. He teaches him the *ladder* construction, a recursive process, which starts with integer pair $x_1 = z_1 = 1$, and then derives subsequent pairs according to the rule

$$x_n + z_n = x_{n+1} \quad \text{and} \quad 2x_n + z_n = z_{n+1}. \tag{4.5}$$

A modern student would understand that his or her Greek counterpart, in his vain attempt to solve the equation $2x^2 - z^2 = 0$, which has no solution with integral values of x and z, was in effect discovering the successive integral values of x for which $z = \sqrt{2x^2 \pm 1}$ is an integer. In other words, the integral solutions of the Diophantine indeterminate equation

$$2x^2 - z^2 \pm 1 = 0. \tag{4.6a}$$

A *Diophantine equation* is one where the coefficients are rational, and whose solutions are also rational.[5] An *indeterminate* system of equations is one where the number of variables exceeds that of distinct equations.

Equation (4.6a) is a particular case of *Pell's equation*, which has received a great deal of attention from number theorists. Pell's equation is written as

$$z^2 - Nx^2 \pm 1 = 0. \tag{4.6b}$$

---

[5]Diophantus was born in Alexandria, during the "Silver Age" of Greek mathematics, ca. A.D. 210. That period was one of renewal, after a near-total decline of ancient mathematics following the death of Archimedes, with a few exceptional figures, such as Appolonius of Alexandria (ca. 180 B.C.), Menelaus of Alexandria (ca. A.D. 100), and Nichomachus of Geresa (ca. A.D. 100). Diophantus published the earliest known treatise of algebra, centuries before the introduction of that discipline by the Arabs. The celebrated *Diophanti Alexandrini Arithmeticorum* was among the last contributions of ancient Greeks to science, and was a principal source of inspiration to Fermat, who recorded his famous last theorem on his personal Latin translation of that book. The *Arithmeticorum* was written nearly half a millenium after the murder of Archimedes by a Roman soldier, and just a few short years before the troops of Septima Zenobia, queen of Palmyra, conquered Egypt in A.D. 269 and completed the burning of the Alexandrian Library.

The reader may verify, using equation (4.5), that if $2x_n^2 - z_n^2 = 1$ for some value of n, then $2x_{n+1}^2 - z_{n+1}^2 = -1$, and if $2x_n^2 - z_n^2 = -1$, then $2x_{n+1}^2 - z_{n+1}^2 = +1$. We have yet to show, however, that no intermediate values of x and z, other than those given by (4.5), satisfy equation (4.6a).

In the ladder shown in Table 4.2a , the ratio z/x offers increasingly better approximations of $\sqrt{2}$. For example, $239/169 \approx 1.4142012$, to be compared, rounded to the same number of decimal digits, to $\sqrt{2} \approx 1,4142136$. The ratios y/x are alternatively larger and smaller than $\sqrt{2}$, as they gradually converge to that value.

It is believed that the ladder procedure allowed Theodorus of Cyrene to also calculate approximate values for the square roots of 3, 5, 7, 8, 10, 11,12, 13, 14, and 15. According to Plato, he stopped short of 17.

A Variation on the Ladder of Theodorus

Suppose that instead of attempting to discover an isosceles right triangle of integral side x, whose hypotenuse z is equal to $\sqrt{2x^2 \pm 1}$ and is an integer, we give ourselves a different challenge, namely, that of discovering a right triangle with integral sides x, $x + 1$ whose hypotenuse is an integer. For our conditions to be met, source integers a, b must satisfy

$$2ab - (a^2 - b^2) = 1, \quad \text{i.e.,} \quad a^2 - 2ab - (b^2 - 1) = 0.$$

We must therefore have $a = b + \sqrt{2b^2 - 1}$, and the problem thus boils down to finding successive values of integer b for which $\sqrt{2b^2 - 1}$ is also an integer. This can be achieved by means of the ladder shown in Table 4.2b, which comprises the odd-rank rungs of the ladder of Table 4.2a. It suggests, by induction, that there exist an infinite number of Pythagorean triples of the form $(x, x + 1, z)$.

The reader's attention is drawn to the relationship between the integers in the above ladders and what I have defined in another book[6] as the Fibonacci sequence of order 2, namely 1, 2, 5, 12, 29, 70, ..., where each term x_i is obtained from the preceding two according to the rule $x_i = 2x_{i-1} + x_{i-2}$.

[6]Midhat J. Gazalé, *Gnomon: From Pharoahs to Fractals* (Princeton N.J.: Princeton University Press, 1999).

TABLE 4.2b
A variation on the Ladder of Theodorus.

b	$\sqrt{2b^2-1}$	a	x	y	z
1	1	2	3	4	5
5	7	12	119	120	169
29	41	70	4 059	4 060	5 741
169	239	408	137 903	137 904	195 025
985	1 393	2 378	4 684 659	4 684 660	6 625 109
5 741	8 119	13 860	159 140 519	159 140 520	225 058 336
. . .					

Table 4.2b generates better and better statements about the value of $\sqrt{2}$. The ladder's fourth rung yields

$$\frac{195\,025}{137\,904} < \sqrt{2} < \frac{195\,025}{137\,903}.$$

The average value of the above two fractions, namely, 1.414213562, exactly corresponds to $\sqrt{2}$ to the tenth decimal.

Fermat's Last Theorem

J'ai découvert une démonstration assez remarquable de cette
proposition, mais elle ne tiendrait pas dans cette marge.
(Pierre de Fermat)

The French mathematician Pierre de Fermat (1601–65) conjectured that 2 is the greatest value of integer n for which there exist integer triples x, y, z satisfying the Diophantine equation

$$x^n + y^n = z^n.$$

Fermat himself, using the method of infinite descent, proved that statement for $n = 3$ and 4. In the margin of his personal copy of the Latin translation by C. G. Bachet of the *Arithmeticorum* of Diophantus, Fermat stated that he had discovered a truly remarkable proof, too long to be contained in the book's margin. The conjecture, which became known as *Fermat's last (or great) theorem*, remained unproven for all other values of n until June 23, 1993, when Andrew Wiles of Princeton University offered a very complex development, which drew upon a vast body of recent mathematical acquisitions. Following minor modifications, his work was finally accepted as a valid proof of Fermat's theorem, and acclaimed by the mathematical community.

It is quite remarkable, and indeed characteristic of arithmetic, that statements as simple as Fermat's theorem would require such incredibly elaborate proofs. Fermat himself, in his correspondence with Mersenne, had claimed in a marginal note, perhaps in earnest, that he had found a simple and elegant proof. The complexity of Wiles's proof inclines one to think that Fermat, with the mathematical tools of his time, could not possibly have arrived at a bona fide proof, and his theorem remained an unproven conjecture for more than three centuries. Excellent articles on the subject have been published by Kenneth A. Ribet and Brian Hayes.[7] Fascinating stories of the proof may also be found in *Fermat's Enigma* by Simon Singh,[8] and *Fermat's Last Theorem*, by Amir Aczel.[9]

The Irrationality of $\sqrt{2}$

The Ladder of Theodorus of Cyrene may have given the Greek student the signal that no matter how far the exercise was stretched, the shortfall $|2x^2 - z^2|$ would never be less than 1. Whereas one could always erect a square upon the hypotenuse of any given right triangle, and declare that its area was equal to the sum of the areas of the squares erected on its sides, it was now impossible to declare that given any two members of the universe of rational numbers, a third member could always be found whose square was equal to the sum of the squares of the other two. Pythagoras's theorem, which was irrefutable in the geometrical realm, could not be transposed to the realm of rational numbers. The universe of rational numbers did not contain all the numbers. Numbers existed that were not rational!

The discovery of the unutterable ($\alpha\lambda o\gamma o\nu$) irrational numbers constituted what the nineteenth-century French mathematician Paul

[7]Kenneth Ribet and Brian Hayes, "Fermat's Last Theorem and Modern Arithmetic," *American Scientist* **82** (March–April 1994): 144–156.

[8]Simon Singh, *Fermat's Enigma* (New York: Walker, 1997).

[9]Amir D. Aczel, *Fermat's Last Theorem* (New York: Four Walls Eight Windows, 1996).

Tannery (noted as co-editor, with Charles Adam, of *Oeuvres de Descartes*) called *un véritable scandale logique*.

A classical proof of the irrationality of $\sqrt{2}$ may be found in Euclid's *Elements*, and we have chosen to illustrate it with the aid of the logical diagram of Figure 4.6, where each proposition follows from the one

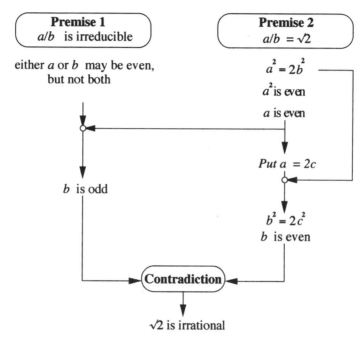

Figure 4.6. Proof of the irrationality of $\sqrt{2}$.

immediately preceding it, or from the conjunction of two preceding propositions. The proof starts with the proposition that the square root of two is a rational number represented by irreducible fraction a/b, and then proceeds by *reductio ad absurdum*.

The following simple proof, which is predicated on the base 3 positional numeration system, may also be offered:

If a base 3 number representation ends in	0	1	2,
then its square ends in	0	1	1,
and twice its square ends in	0	2	2.

Integers a and b may satisfy $a^2 = 2b^2$ only if a digit on the second line is equal to the opposite digit on the third line, and that happens only for digit 0. In other words, both a and b must be multiples of 3, which is not an acceptable hypothesis, since a/b must be irreducible.

That proof is predicated upon Fermat's *little theorem*, which states that for any prime p and any integer a not a multiple of p, we get $a^{p-1} \equiv 1 \pmod{p}$. It follows that for any integer a not a multiple of 3, we get $a^2 \equiv 1 \pmod{3}$. The congruence $a^2 \equiv 2b^2 \pmod{3}$ therefore has no solution unless a and b are both divisible by 3.

A (Theoretically) Physical Impossibility

Figure 4.7a illustrates a simple experiment in which a weightless box containing five identical marbles is allowed to slide without friction down a slope of 3/4. An ideal string wrapped around a frictionless pulley is attached to the box at one extremity, and to a weightless tray at the other, upon which marbles may be placed, identical to those in the box. Elementary mechanics teaches us that if the vertical component of the box's weight is 5 units, the downward force parallel to the inclined plane is 3 units. It is therefore sufficient to place three marbles on the tray in order to attain equilibrium. In Figure 4.7b, the inclined plane's slope is 1/1. If the box contains M marbles, no integral number M' of identical marbles may be found which balances the

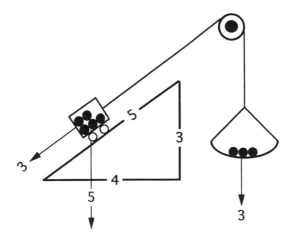

Figure 4.7a. Commensurable weights.

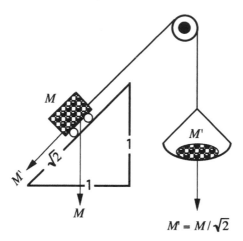

Figure 4.7b. Incommensurable weights.

box when placed on the tray. That number would have to be equal to $M/\sqrt{2}$ in order to achieve equilibrium, and that is impossible. M and M' are incommensurable.

The square root of integer 2 is not a rational number, and was therefore declared *irrational*. But does that exclusion constitute a definition of irrationality?

DEDEKIND

> Just as negative and rational numbers are formed by a new
> creation, and as the laws of operating with these numbers
> must and can be reduced to the laws of operating with
> positive integers, so we must endeavor completely to define
> irrational numbers by means of the rational numbers alone.
> (Dedekind)

Having understood commensurability and rational numbers, we now legitimately ask ourselves if we fully understand what irrationality really means.

A definition of the equality of two irrationals was offered around 360 B.C. by a genius named Eudoxus of Cnidus, miraculously ahead of the paradigms of his time, and it was not until 1872 that a definition of irrationals was offered by an otherwise obscure mathematician, which definition was eventually accepted by the mathematical establishment, and prevails to this day.

The definition of irrationals may be stated in surprisingly simple terms, as follows: If the universe of number is partitioned into two distinct classes L and H (L for low, and H for high), by means of a *cut*, so that every rational member of L is less than every rational member of H, and no number may belong to both L and H, there is one and only one number that can achieve that partition. If L has a largest rational *or* if H has a smallest rational, the cut is that rational number. If L has no largest rational *and* H has no smallest rational, the cut is an irrational number.

That definition of an irrational number as a *Schnitt*, or partitioning of rational numbers, was offered in 1872 by the German mathematician Richard Dedekind (1831–1916). Just as rational numbers were defined, from within, in terms of integers, so were irrational numbers defined, also from within, in terms of rational numbers.

The fundamental theorems on limits, which constitute the cornerstone of analysis, could thence receive rigorous proofs without recourse to geometry. In the words of Carl B. Boyer and Uta C. Merzbach:

> It was geometry that had pointed out the way to a suitable definition of continuity, but in the end it was excluded from the formal arithmetic definition of the concept. The Dedekind cut in the rational number system, or an equivalent construction of real number, now has replaced geometrical magnitude as the backbone of analysis.[10]

The set of rational numbers and that of irrational numbers together constitute the continuum of real numbers. The continuum of real numbers consists of those numbers that may be treated as the ratio of two integers, namely, rationals, and the others, namely, irrationals. The latter in turn consist of two subsets, that of algebraic irrationals, that is, the roots of polynomial equations with rational coefficients, and the others, referred to as *transcendental irrationals*. It is the continuum of numbers, not the the set of rationals alone, that is gapless, as spelled out by the Cantor-Dedekind axiom, which postulates that a one-to-one correspondence exists between the points on

---

[10]Carl B. Boyer and Uta C. Merzbach, *A History of Mathematics* (New York: Wiley, 1991).

a line (in our case, the lines on a plane) and the continuum of *real numbers.*

Again quoting E. C. Titchmarsh:

> It might be thought for an awful moment that, if we operated again with irrational numbers as we did with rational numbers, we should come upon a new class of super-irrationals, and so on endlessly. But this is not so. Nothing fresh emerges from this process. The irrational numbers already defined are the only ones which exist.[11]

Unfortunately, that is no longer true, and John Conway has shown that a very natural simplification of Dedekind's *Schnitt* generates many more numbers that fall in neither category. For an elementary introduction to Conway's *Surreal Numbers*, see Knuth's *Surreal Numbers*, and Conway and Guy's *Book of Numbers.*[12]

EUDOXUS

Willingly would I burn to death like Phaeton,
were this the price for reaching the sun and
learning its shape, its size, and its substance.
(Eudoxus)

Eudoxus is rightly regarded as one of the greatest mathematicians, in addition to being an outstanding astronomer. Educated in the Academy, he inspired Ptolemy, more than half a millenium later, to build the geocentric system, which remained unchallenged until Copernicus, in 1543. His contribution to alleviating the almost insurmountable ambiguities that shrouded irrational numbers had an extraordinarily modern quality. Said James R. Newman, "For more than 2000 years, Eudoxus' definition provided the only basis for handling irrational numbers." It is on the very foundation of the magnificent definition given by Eudoxus of Cnidus (408–355 B.C.) that Dedekind established, in 1872, the legitimacy of the *Schnitt*, or partition of rational numbers into two classes by an irrational number, as the very definition of that irrational number.

---

[11] E. C. Titchmarsh, *Mathematics for the General Reader* (New York: Dover, 1981).
[12] Donald E. Knuth, *Surreal Numbers* (1974); John H. Conway and Richard K. Guy, *The Book of Numbers* (New York: Springer Verlag, 1996).

The definition of Eudoxus, as it appeared in Book V of Euclid's *Elements*, reads as follows:

> Magnitudes are said to be in the same ratio, the first to the second and the third to the fourth, when, if any equimultiples whatever be taken of the first and the third, and any equimultiples whatever of the second and the fourth, the former equimultiples alike exceed, are alike equal to, or are alike less than, the latter equimultiples taken in corresponding order.[13]

In order to interpret that statement, we shall resort to a mechanical metaphor, the lever, which Archimedes extensively used in his study of volumes and areas, and upon which he founded a veritable intellectual method. Using the notion of the lever, we shall perform a *gedanken-experiment*, or experiment of the mind, which will hopefully shed more light on the nature of irrational numbers.

The lever may be viewed as an idealized straight weightless bar, articulated somewhere along its length on an *infinitely sharp* fulcrum. Identical weightless trays hang vertically from its extremities, upon which various weights can be placed, the ultimate object being to reach static equilibrium. The observations of Archimedes led him to articulate his law of levers, which states that arbitrary magnitudes placed at the lever's extremities balance each other when their weights are inversely proportional to their respective distance from the fulcrum.[14]

Imagine that a lever is divided into a section a measuring 3 units of length left of the fulcrum, and a section b measuring 5 units to its right. Imagine also that we possess a large quantity of little beads of equal weight, and that we wish to balance the scales by placing an appropriate number of beads on each tray. According to the law of the lever, if we place m beads on the left tray and n beads on the right tray, equilibrium is achieved whenever $3m = 5n$, in other

---

[13]See T. L. Heath, *The Thirteen Books of Euclid's Elements* (Cambridge, 1908).

[14]That discovery led him to announce to the world that, given a sufficiently long and strong lever, resting on an immutable fulcrum, he could lift the earth. Short of achieving that feat, Archimedes nonetheless conceived redoubtable war machines, which spread terror amongst Roman ranks, whether they were actually built or not. His reputation eventually caused him to be slain by a Roman soldier, as he was tracing a geometric figure in the dust, following the fall of Syracuse.

words, for any pair m, n such that $m/n = 5/3$. There exists an infinite number of such pairs, all of which balance the scales. Since m and n are integers, equilibrium can be achieved for arbitrary lengths a, b if their magnitudes are commensurable, in other words, if a/b is rational.

Now imagine that a very able craftsman possesses a perfectly square template that allows him to manufacture a pair of scales such that a is equal to the square's side, and b to its diagonal. He now attempts to balance the scales by progressively adding identical little beads on one or the other tray, depending on which side tips. Figure 4.8 shows a sampling of of near-equilibrium stages, where the scales initially tip one way, and upon the addition of a single bead on the opposite tray, they tip the other way. Arrows indicate which side tips. The critical stages in the figure can be seen to correspond to the rungs of the Ladder of Theodorus.

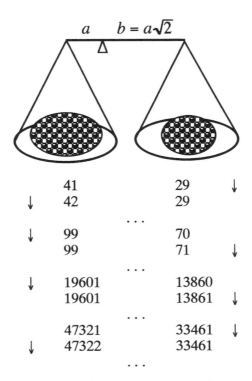

Figure 4.8. A *gedanken*-experiment with scales.

No matter how far out we proceed, we repeatedly fall back on similar critical circumstances, only a fraction of a bead away from equilibrium. As the beads cannot be fragmented, equilibrium is never reached. The lever thus affords a nice physical metaphor for incommensurability, akin to the inclined plane.

The next step in our *gedanken*-experiment requires two levers of different lengths, of which we are told that they are divided according to the same proportion by their respective fulcrums. We are asked to experimentally ascertain that it is really so, using as many beads as we wish.

Let the first lever be of lengths a and b on either side of its fulcrum. Let the corresponding lengths of the second lever be c and d.

If a pair m, n is found to balance the first lever as well as the second, we deduce that $a/b = c/d$ is a rational number. That is equivalent to saying that if there exists a pair m, n such that $ma = nb$ and $mc = nd$, then $a/b = c/d$. It follows that any integer pair m', n' that balances one lever must also balance the other.

If a/b and c/d are not rational, we have just seen that it is impossible to balance the levers, no matter how large a stock of beads we have at our disposal. We may only observe that the scales tip to one side or the other. We are authorized, however, to conduct an *infinite* number of trials. If we observe that every pair of values m, n that tips the first lever to one side also tips the second to the same side, then, according to Eudoxus, it may be declared that $a/b = c/d$. In other words, if for every pair m, n such that $ma < nb$, we also have $mc < nd$, or if for every pair m, n such that $ma > nb$, we also have $mc > nd$, then $a/b = c/d$. The definition given by Eudoxus may thus be rephrased as follows:

> The ratio of a to b is equal to the ratio of c to d if, whatever multiples ma and mc are chosen, and whatever multiples nb and nd are chosen, then
> either $ma > nb$ and $mc > nd$;
> or $ma = nb$ and $mc = nd$;
> or $ma < nb$ and $mc < nd$.

The above definition is generally known as the axiom of Archimedes, though the great Archimedes himself gave credit for it to Eudoxus, half a millenium later.

MARGINALIA

Three Ancient Problems

Let no one destitute of geometry enter my door.
(Plato)

Around the fourth century B.C., the center of intellectual life shifted to Athens, following the victories of the Greeks over the Persians. Plato (429–348 B.C.) remains one of the most revered philosophers of antiquity. Around 390 B.C., he founded the Academy, named after the Greek hero Academos. Above its entrance were engraved the words "Let no one destitute of geometry enter my door."

The Athenian school made innumerable attempts at solving three major problems of geometry:

i. *The squaring of the circle* consists of tracing a square whose perimeter is equal to that of a given circle.

ii. *The doubling of the cube*, otherwise known as *the Delian problem*, consists of determining the side of a cube whose volume is twice that of a given cube. It is said to have originated when the Greeks appealed to the oracle of Delos (hence its name), following the plague that took the life of Pericles in 429 B.C. Apollo responded by challenging them to double the volume of his cubic altar.

iii. *The trisecting of an angle* consists of dividing a given angle into three equal parts.

Plato stipulated that the challenges had to be met without resorting to any instrument other than an unmarked ruler and a compass (with the additional constraint that the compass not be lifted from the surface upon which the figure was being traced.). These are referred to as the *Platonic restrictions*. The Greeks were never able to solve these problems, and Pappus declared in A.D. 320, almost 750 years after the problems were first enunciated, that they could never be solved, though he was not able to prove why that was so. The problems were proven insoluble fifteen centuries later. The search, however, led en route to substantial early mathematical discoveries. Hippocrates of Chios (not to be confused with the founder of medicine, Hippocrates of Cos) showed that the ratio between the areas of two circles is the same as that between the squares of their diameters, thus bringing to light the existence of some universal constant π yet to be discovered. In his search for that constant, it is believed that Antiphon invented the method of *exhaustion*. Squaring the circle, whether the method used is geometrical or arithmetical, involves the constants π and $\sqrt{2}$. Both numbers are irrational, but only $\sqrt{2}$ is algebraic, that is, the root of an algebraic polynomial equation with integral coefficients. (More precisely, it is a quadratic irrational.) Constructing $\sqrt{2}$ with ruler and compass is quite straightforward: it is the hypotenuse of an isosceles right triangle of side equal to 1. Constructing π, on the other hand, is an entirely different matter. From the work of Ferdinand Lindemann in 1882, we know today that π is *transcendental*; that is, it is not algebraic. It can be shown that such numbers cannot be constructed as prescribed, but that is beyond the scope of the present discussion.

Doubling the cube involves only the algebraic irrational $\sqrt[3]{2}$. The solution cannot be constructed on a plane if the Platonic restrictions are to be abided by, but appropriate geometric constructions exist in three-dimensional space. Archytas of Tarentum (Italy) offered a brilliant solution invoving three intersecting forms, namely, the cylinder, the cone, and the torus. Menaechmus, to whom the solution is sometimes attributed, is credited with discovering the conic sections: the parabola, the hyperbola, and the ellipse. A similarly unorthodox solution for trisecting an angle was proposed by Archimedes, by drawing tangents to the spiral that bears his name.

APPENDIX

Proof of the Irrationality of e

The number e is defined as the infinite series

$$e = 1 + \frac{1}{2} + \frac{1}{2!} + \frac{1}{3!} + \cdots.$$

That series adds up to an irrational number, when carried to infinity. The irrationality of e is rather easy to establish.

Like so many arguments of arithmetic, the following proof, which was found in E. C. Titchmarsh's *Mathematics for the General Reader* proceeds by reductio ad absurdum. Assume e is a rational number, meaning that there exist integers p and q such that

$$\frac{p}{q} = 1 + \frac{1}{1} + \frac{1}{2!} + \frac{1}{3!} + \cdots \qquad (4.7)$$

This may be written as

$$\frac{p}{q} = \left(1 + \frac{1}{1} + \frac{1}{2!} + \frac{1}{3!} + \cdots + \frac{1}{q!}\right) + \left(\frac{1}{(q+1)!} + \frac{1}{(q+2)!} + \cdots\right).$$

We now multiply both sides by $q!$, which is quite legitimate, as the series between the second fenced pair is convergent. We obtain

$$p(q-1)! = \left(q! + \frac{q!}{1} + \frac{q!}{2!} + \frac{q!}{3!} + \cdots + \frac{q!}{q!}\right)$$

$$+ \left(\frac{1}{(q+1)} + \frac{1}{(q+1)(q+2)}\right.$$

$$\left. + \frac{1}{(q+1)(q+2)(q+3)\cdots} + \cdots\right) \qquad (4.8)$$

Both sides of the equality sign on the upper line of (4.8) are integers, suggesting that the expression on the lower lines is also an integer. Resorting to a stratagem similar to that of Nicolas Oresme, we shall compare that line, term for term, with the series

$$S = \left(\frac{1}{(q+1)} + \frac{1}{(q+1)(q+1)} + \frac{1}{(q+1)(q+1)(q+1)\cdots} + \cdots\right).$$

$$(4.9)$$

Clearly, series S adds up to a larger number than the lower lines of (4.8). On the other hand, we have already seen that for any positive number x,

$$\frac{1}{x} + \frac{1}{x^2} + \frac{1}{x^3} + \cdots = \frac{1}{x-1}.$$

It follows that

$$S = \frac{1}{q}.$$

S is a fractional number, and that number is larger than the lower part of (4.8) which is supposed to be an integer. We are in the presence of a contradiction, and our hypothesis, according to which e was a rational number, fails.

CHAPTER 5

Continued Fractions

Continued fractions are part of the "lost mathematics,"
the mathematics now considered too advanced
for high school and too elementary for college.
(Petr Beckman)[1]

This chapter dwells on a special kind of fraction, which is widely used to calculate the values of algebraic irrationals, such as surds, as well as the values of certain transcendental irrationals, such as e or π. It is believed that continued fractions were first introduced by William Brouncker (1620–1684), the first president of the British Royal Society, who discovered a beautiful expression for the transcendental number π, to which we shall soon return.

By merely looking at a continued fraction, one is struck by its iterative character, which appears at once. Whereas the positional representation of quadratic irrationals with respect to a periodic base, be it decimal or otherwise, is not periodic, the corresponding continued fraction is periodic, and may thus be defined in terms of a finite number of elements. Similarly, the continued fractions representing e and π obey simple, albeit nonperiodic, patterns. These patterns are epitomized by Euler's continued fractions for e.

EUCLID'S ALGORITHM

As an introduction to continued fractions, we shall first examine a celebrated ancient Greek algorithm. That algorithm is found in Book VII of Euclid's *Elements*, and is intended to calculate the greatest common divisor of any two integers. Though it is ascribed by several historians to the great Eudoxus, it is usually referred to as Euclid's algorithm.

[1]Petr Beckman, *A History of π* (New York: St. Martin's Press, 1971).

Consider integers a, b, with $b > 0$. It was established in chapter 2 that there exists one and only one integer pair q, r such that

$$a = bq + r \qquad b > r \geq 0. \tag{5.1}$$

Let d be an integer that divides both a and b, and write

$$\alpha = a/d \qquad \beta = b/d. \tag{5.2a}$$

Substituting in (5.1), we get

$$\alpha = \beta q + r/d. \tag{5.2b}$$

In the above equation, both α and βq are integers. Consequently, r/d is also an integer, signifying that if d divides both a and b, it also divides r, the remainder of the division of a by b. Conversely, it is clear that any common divisor of b and r is also a divisor of a. The set of common divisors of a and b is therefore identical to the set of common divisors of b and r. The largest number in this set is the greatest common divisor (GCD) of a and b, which is also the GCD of b and r. This is written

$$(a, b) = (b, r). \tag{5.3}$$

That simple property of division is the foundation upon which Euclid based his algorithm, which proceeds as follows: It is required to determine the GCD of integers a_0 and a_1, where $a_0 > a_1$. We may write, following the pattern of equation (5.1),

$$\begin{aligned}
a_0 &= a_1 q_0 + a_2 & a_1 &> a_2 \\
a_1 &= a_2 q_1 + a_3 & a_2 &> a_3 \\
a_2 &= a_3 q_2 + a_4 & a_3 &> a_4 \\
a_3 &= a_4 q_3 + a_5 & a_4 &> a_5,
\end{aligned} \tag{5.4}$$

$$\cdots$$

$$a_{i-1} = a_i q_{i-1} + a_{i+1} \qquad a_i > a_{i+1}$$

$$\cdots$$

$$a_{n-1} = a_n q_{n-1} + 0.$$

Residue $a_n = 0$ is bound to eventually occur for some value n, since integer sequence $a_0 > a_1 > a_2 > a_3 \cdots > a_n$ can contain no more than a_0 *positive decreasing integers*. Its occurrence signals that a_n is the GCD of a_{n-1} and itself, in other words, that $(a_{n-1}, a_n) = a_n$. From the above sequence of divisions, we have, according to equation (5.3),

$$(a_0, a_1) = (a_1, a_2) = (a_2, a_3) = \cdots = (a_{n-1}, a_n).$$

The sought-after GCD is therefore a_n.

For example, to determine the GCD of 1785 and 374:

$$1785 = 374 \times 4 + 289,$$
$$374 = 289 \times 1 + 85,$$
$$289 = 85 \times 3 + 34,$$
$$85 = 34 \times 2 + 17,$$
$$34 = 17 \times 2 + 0.$$

Hence, $(1785, 374) = 17$.

CONTINUED FRACTIONS

Putting $\phi_0 = \dfrac{a_0}{a_1}$, $\phi_1 = \dfrac{a_1}{a_2}$, $\phi_2 = \dfrac{a_2}{a_3}$, $\ldots$, $\phi_{n-1} = \dfrac{a_{n-1}}{a_n}$, equation (5.4) may be written as

$$\phi_0 = \frac{a_0}{a_1} = q_0 + \frac{a_2}{a_1} = q_0 + \frac{1}{\phi_1},$$

$$\phi_1 = \frac{a_1}{a_2} = q_1 + \frac{a_3}{a_2} = q_1 + \frac{1}{\phi_2},$$

$$\cdots \tag{5.5}$$

$$\phi_{n-2} = \frac{a_{n-2}}{a_{n-1}} = q_{n-2} + \frac{a_n}{a_{n-1}} = q_{n-2} + \frac{1}{\phi_{n-1}},$$

$$\phi_{n-1} = \frac{a_{n-1}}{a_n} = q_{n-1} + 0,$$

which in turn can be written as

$$\phi_0 = \frac{a_0}{a_1} = q_0 + \cfrac{1}{q_1 + \cfrac{1}{q_2 + \cfrac{1}{q_3 + \cdots}}}$$

(5.6a)

$$\cdots q_{n-2} + \cfrac{1}{q_{n-1}}.$$

This is called a *continued fraction*, and $q_0, q_1, q_2, \ldots$ are called the *partial quotients*.

For the previous numerical example, following the preceding pattern, we may write

$$\phi_0 = \frac{1785}{374} = 4 + \frac{1}{374/289} \qquad q_0 = 4,$$

$$\phi_1 = \frac{374}{289} = 1 + \frac{1}{289/85} \qquad q_1 = 1,$$

$$\phi_2 = \frac{289}{85} = 3 + \frac{1}{85/34} \qquad q_2 = 3, \qquad (5.6b)$$

$$\phi_3 = \frac{85}{34} = 2 + 34/17 \qquad q_3 = 2,$$

$$\phi_4 = \frac{34}{17} = 2 \qquad q_4 = 2,$$

and we get

$$\phi_0 = \frac{1785}{374} = 4 + \cfrac{1}{1 + \cfrac{1}{3 + \cfrac{1}{2 + \cfrac{1}{2}}}}$$

(5.6c)

Regular Continued Fractions

The previous discussion suggests the following general form for continued fractions:

$$\phi = q_0 + \cfrac{p_1}{q_1 + \cfrac{p_2}{q_2 + \cfrac{p_3}{q_3 + \cdots}}} \tag{5.7}$$

$$q_{n-1} + \frac{p_n}{q_n}.$$

For two striking examples, see expressions (5.16) and (5.17).

A particular case results from putting $p_i = 1$ for all i, which corresponds to what is called a *simple continued fraction* (SCF). If, additionally, all qs are positive integers, the SCF is said to be a *regular continued fraction* (RCF). The following notation will be used for regular continued fractions:

$$[q_0, q_1, q_2, \ldots, q_n] = q_0 + \cfrac{1}{q_1 + \cfrac{1}{q_2 + \cfrac{1}{q_3 + \cdots}}} \tag{5.8}$$

$$q_{n-1} + \frac{1}{q_n},$$

and we may easily show that

$$[q_0, q_1, q_2, \ldots, q_n] = [q_0, [q_1, q_2, \ldots, q_n]]$$

$$= q_0 + \frac{1}{[q_1, q_2, \ldots, q_n]}, \tag{5.9}$$

$$a[q_0, q_1, q_2, \ldots] = \left[aq_0, \frac{q_1}{a}, aq_2, \frac{q_3}{a}, \ldots\right].$$

Convergents

The irreducible fraction

$$\delta_i = \left[q_0, q_1, q_2, \ldots, q_i\right] = \frac{N_i}{D_i} \tag{5.10}$$

is called the continued fraction's ith *convergent*. N_i and D_i are, respectively, the fraction's ith numerator and denominator.

Returning to the numerical example, we get

$$\delta_0 = 4,$$

$$\delta_1 = 4 + \frac{1}{1} = 5,$$

$$\delta_2 = 4 + \cfrac{1}{1 + \cfrac{1}{3}} = \frac{19}{4},$$

$$\delta_3 = 4 + \cfrac{1}{1 + \cfrac{1}{3 + \cfrac{1}{2}}} = \frac{49}{9},$$

$$\delta_4 = 4 + \cfrac{1}{1 + \cfrac{1}{3 + \cfrac{1}{2 + \cfrac{1}{2}}}} = \frac{105}{22} = \frac{1785}{374}.$$

If, for consistency, we introduce "virtual" numerators and denominators,

$$N_{-1} = 1, \qquad N_{-2} = 0, \qquad D_{-1} = 0, \qquad D_{-2} = 1, \tag{5.11}$$

we may derive two fundamental recursion formulae, which, together, yield the successive irreducible convergents:

$$N_i = N_{i-2} + q_i N_{i-1} \quad \text{and} \quad D_i = D_{i-2} + q_i D_{i-1}. \tag{5.12}$$

Hence the procedure illustrated by Table 5.1.

TABLE 5.1
Procedure for Generating Convergents.

i	-2	-1	0	1	2	3	4
q_i	–	–	4	1	3	2	2
N_i	0	1	4	5	19	43	105
D_i	1	0	1	1	4	9	22
∂_i	–	–	4	5	4.75	4.7	4.7$\underline{72}$

Huygens showed that in the general case,

$$N_{i-1}D_i - N_iD_{i-1} = (-1)^i. \qquad (5.13)$$

Terminating Regular Continued Fractions

Let us return to development (5.5), where the number ϕ_0 may or may not be an integer, and write

$$\phi_0 = q_0 + \frac{1}{\phi_1}, \quad \text{with } \phi_1 > 1,$$

$$\phi_1 = q_1 + \frac{1}{\phi_2}, \quad \text{with } \phi_2 > 1,$$

$$\phi_2 = q_2 + \frac{1}{\phi_3}, \quad \text{with } \phi_3 > 1,$$

etc.

Index s is reached, for which $\phi_s = q_s$ is an integer. The process halts, and we are in presence of the *terminating RCF*,

$$\phi_0 = [q_0, q_1, q_2, \ldots, q_s],$$

signaling that ϕ_0 is a rational number. Every terminating RCF thus generates a rational number and, by applying Euclid's algorithm, every rational number can be generated by an RCF.

Quotient q_s is called the *terminating quotient*, or *termination*.

Periodic Regular Continued Fractions

Indices s and t are reached, such that $\phi_{s+t} = \phi_s$ and s is the lowest index for which that occurs. The SCF is said to be periodic, of lead length s and period t.

Let us consider a few examples. As our first example, to find the RCF corresponding to $\sqrt{14}$, we write the following.[2]

$$
\begin{aligned}
\phi_0 &= \sqrt{14} &&= 3 + \left(\sqrt{14} - 3\right) \\
\phi_1 &= \frac{1}{\sqrt{14} - 3} &&= 1 + \frac{\sqrt{14} - 2}{5} \\
\phi_2 &= \frac{5}{\sqrt{14} - 2} &&= 2 + \frac{\sqrt{14} - 2}{2} \\
\phi_3 &= \frac{2}{\sqrt{14} - 2} &&= 1 + \frac{\sqrt{14} - 3}{5} \\
\phi_4 &= \frac{5}{\sqrt{14} - 3} &&= 6 + \left(\sqrt{14} - 3\right) \\
\phi_5 &= \frac{1}{\sqrt{14} - 3} &&
\end{aligned}
\tag{5.14}
$$

In this example, $\phi_5 = \phi_1$, meaning that the resulting RCF is of lead length 1 and period 4, i.e., $\sqrt{14} = [3, 1, 2, 1, 6, 1, 2, 1, 6, 1, 2, \ldots]$ $= [3, \underline{1, 2, 1, 6}]$.

An easy example is supplied by $\sqrt{2}$. Given that $2\sqrt{2} + 3 = (\sqrt{2} + 1)^2$, we may write

$$
\phi_0 = \sqrt{2} + 1 = \frac{2\sqrt{2} + 3}{\sqrt{2} + 1} = 2 + \frac{1}{\sqrt{2} + 1} = 2 + \frac{1}{\phi_0}.
$$

Hence,

$$
\sqrt{2} + 1 = [2, 2, 2, 2, \ldots],
$$

and we obtain

$$
\sqrt{2} = [1, 2, 2, 2, \ldots] = [1, \underline{2}].
$$

---

[2] The method used for generating the identities in the example is described in the appendix to this chapter.

The underlining scheme signifies that integer 2 is indefinitely repeated. Using the tabular method to calculate the successive convergents of $\sqrt{2}$, we get the values shown in Table 5.2a. The successive convergents oscillate around asymptotic value $\sqrt{2}$, as shown in Table 5.2b. That pattern, which is typical of SCFs, is reminiscent of the

TABLE 5.2a
Calculating the convergents of $\sqrt{2}$.

i	−2	−1	0	1	2	3	4	5	6
q_i	–	–	1	2	2	2	2	2	2
N_i	0	1	1	3	7	17	41	99	239
D_i	1	0	1	2	5	12	29	70	169
∂_i	–	–	$\dfrac{1}{1}$	$\dfrac{3}{2}$	$\dfrac{7}{5}$	$\dfrac{17}{12}$	$\dfrac{41}{29}$	$\dfrac{99}{70}$	$\dfrac{239}{169}$

TABLE 5.2b
Convergents of $\sqrt{2}$.

	1.5		1.41\underline{6}		1.4142857...			...
	↑	↓	↑	↓		↑	↓	↑
1		1.4		1.413793...			1.4142012...	

damped oscillations so familiar to electrical engineers. In the words of H. W. Turnbull,

> As these successive ratios are alternately less than and greater than all that follow, they nip the elusive limiting ratio between two extremes, like the ends of a closing pair of pincers. They approximate from both sides to the desired irrational. Like pendulum swings of an exhausted clock, they die down—but they never actually come to rest.[3]

[3]Herbert Western Turnbull, "The Great Mathematicians," in *The World of Mathematics*, edited by James R. Newman (New York: Simon and Schuster, 1956), p. 97.

TABLE 5.3

Generating the convergents of $\phi = (1 + \sqrt{5})/2$.

i	-2	-1	0	1	2	3	4	5	6
q_i	–	–	1	1	1	1	1	1	1
N_i	0	1	1	2	3	5	8	13	21
D_i	1	0	1	1	2	3	5	8	13
∂_i	–	–	$\dfrac{1}{1}$	$\dfrac{2}{1}$	$\dfrac{3}{2}$	$\dfrac{5}{3}$	$\dfrac{8}{5}$	$\dfrac{13}{8}$	$\dfrac{21}{13}$

This example illustrates the use of continued fractions for the approximation of quadratic irrational numbers. A reasonably good and easy to remember approximation of $\sqrt{2}$ is provided by the convergent ∂_5 in Table 5.2a, namely, 99/70. That is exactly the height-width ratio of a standard sheet of stationery in France, whose sides measure 29.7 by 21 centimeters, a remarkable choice whose merit is that no matter how many times the sheet is folded around its shorter middle line, the resulting sheet is geometrically similar to the original.

A further example is provided by the often cited quadratic number known as the Golden Section, namely, $\phi = (1 + \sqrt{5})/2 = [1, 1, 1, 1, \ldots]$. Table 5.3 shows its first seven convergents.

It may be shown that a periodic RCF always converges to a quadratic algebraic irrational, in other words, a number of the form $(a + \sqrt{b})/c$, where a, b, and c are integers and b is not a perfect square. The French mathematician Lagrange proved the converse of that statement, namely, that any quadratic number can be expressed in the form of a periodic SCF.

Spectra of Surds

Early Arabic mathematical textbooks referred to square roots as *samet*, meaning "mute," which was transposed to *surd* in the original translations from the Arabic. A surd's continued fraction is always of the form

$$\sqrt{N} = [\alpha, \beta, \chi, \delta, \ldots \delta, \chi, \beta, \omega],$$

where $\omega = 2\alpha$, as in $\sqrt{2} = [1, \underline{2}]$, $\sqrt{3} = [1, \underline{1, 2}]$, $\sqrt{5} = [2, \underline{4}]$, $\sqrt{14} = [3, \underline{1, 2, 1, 6}]$.

Tobias Dantzig referred to the integers between square brackets as the surd's *spectrum*. Table 5.4 gives the spectra corresponding to a few surds.

The RCFs are periodic, following a *lead sequence* of length 1, consisting of integer α, which is the integral part of the surd.

TABLE 5.4
Spectra of surds.

N	α		ω
2	1		2
3	1	1	2
5	2		4
6	2	2	4
7	2	1 1 1	4
8	2	1	4
10	3		6
11	3	3	6
12	3	2	6
13	3	1 1 1 1	6
14	3	1 2 1	6
15	3	1	6
17	4		8
18	4	4	8
19	4	2 1 3 1 2	8
20	4	2	8
21	4	1 1 2 1 1	8
22	4	1 2 4 2 1	8
23	4	1 3 1	8
24	4	1	8

Nonperiodic, Nonterminating Regular Continued Fractions

If the RCF is neither terminated nor periodic, that situation corresponds to algebraic numbers of degree higher than 2, and to transcendental (nonalgebraic) irrationals, such as e, for which Euler (1707–1783) himself discovered the representation

$$e = [2, 1, 2, 1, 1, 4, 1, 1, 6, 1, 1, 8, \ldots].$$ (5.15)

The convergence pattern shown in Table 5.5 is similar to that observed in the case of periodic continued fractions: the number is approached by successive ratios, which are alternately less and greater than all that follow. With ten-decimal-digit precision, convergent $\partial_6 \sim 2{,}717948718\ldots$ is only $.012$ percent removed from e.

TABLE 5.5
Generating the convergents of e.

i	-2	-1	0	1	2	3	4	5	6
q_i	–	–	2	1	2	1	1	4	1
N_i	0	1	2	3	8	11	19	87	106
D_i	1	0	1	1	3	4	7	32	39
∂_i	–	–	$\dfrac{2}{1}$	$\dfrac{3}{1}$	$\dfrac{8}{3}$	$\dfrac{11}{4}$	$\dfrac{19}{7}$	$\dfrac{87}{32}$	$\dfrac{106}{39}$

It may be shown that every nonterminating RCF is convergent, and that every real number can be expanded in exactly one way as an RCF.

If we regard a number's expansion into an RCF as a "representation" of that number within the "system" of continued fractions, we are struck by the similarities to—and differences from—the positional number system. Both representations are always convergent, and uniquely correspond to a number. Conversely, every number may be uniquely represented in either system. In both cases, terminating expansions represent rational numbers.

Whereas infinite periodic representations correspond to rational numbers in the positional system, they correspond to quadratic irrationals in the RCF system. Transcendental irrationals are obviously represented by infinite nonperiodic representations in both cases.

Two Celebrated Irregular Continued Fractions

William Brouncker discovered the following beautiful expression for the transcendental number π:

$$\frac{4}{\pi} = 1 + \cfrac{1}{2 + \cfrac{9}{2 + \cfrac{25}{2 + \cfrac{49}{2 + \cdots}}}} \qquad (5.16)$$

Euler made substantial contributions to the study of continued fractions, and discovered a no less beautiful expression for e, the other famous transcendental number:

$$e = 2 + \cfrac{1}{1 + \cfrac{1}{2 + \cfrac{2}{3 + \cfrac{3}{4 + \cdots}}}} \qquad (5.17)$$

APPENDIX

How do we get to the statement

$$\frac{5}{\sqrt{14} - 2} = 2 + \frac{\sqrt{14} - 2}{2} \; ?$$

1. We put $x = \dfrac{5}{\sqrt{14} - 2}$.
2. Multiplying the numerator and denominator by $\sqrt{14} + 2$, we get

$$x = \frac{5(\sqrt{14} + 2)}{10} = \frac{\sqrt{14} + 2}{2}.$$

Hence $2x - 2 = \sqrt{14}$, which yields

$$(x - 1)^2 = \frac{14}{4}.$$

We now plug in successive values of $x = 1, 2, 3 \ldots$ until one such value makes the left-hand side larger than the right-hand side:

$$x = 1 \rightarrow (1 - 1)^2 = 0,$$

$$x = 2 \rightarrow (2 - 1)^2 = 1,$$

$$x = 3 \rightarrow (3 - 1)^2 = 4 > \frac{14}{4}.$$

The largest value of x that causes the left-hand side not to exceed the right-hand side is 2. That value is the integral part of x.

3. This allows us to write

$$x = \frac{\sqrt{14} + 2}{2} = 2 + y,$$

where $y < 1$ is the fractional part of x. Solving for y, we get

$$y = \frac{\sqrt{14} - 2}{2},$$

and we obtain

$$\frac{5}{\sqrt{14} - 2} = 2 + \frac{\sqrt{14} - 2}{2}.$$

CHAPTER 6

Cleavages

In this chapter, we introduce a geometric metaphor, which will hopefully give the reader an intuitive "feel" for rationals, irrationals, and infinite processes, as nothing better than a "feel" will ever be possible, even to first-rate mathematicians, which this author confesses not to be. The system was inspired by an artifact that Tobias Danzig briefly resorted to, aimed at geometrically demonstrating the denumerability of rationals, a notion that we shall shortly address.[1] It relies on the use of a pair of orthogonal axes, akin to those of the Cartesian coordinate system, which was invented by the French mathematician and philosopher René Descartes.[2]

The system we are about to introduce is, in fact, more akin to the diagram invented by the Swiss mathematician Jean-Robert Argand, in which complex numbers are represented by points on the plane.[3] That scheme is referred to as the *Argand diagram*, or simply the *complex plane*. In it, the complex number $x + jy$ is represented by a point whose abscissa is x, and whose ordinate is y. Simple as that construction may be, it nonetheless constitutes a very powerful tool in the hands of, among other people, electrical engineers, as it helps them deal with the otherwise elusive sinusoidal variables upon whose behavior their art is predicated.

---

[1] Tobias Dantzig, *Number, the Language of Science*, 4th ed. (Garden City, N.Y.: Doubleday, 1954).

[2] Born in La Haye, Descartes dreamt on November 10, 1619, at the age of twenty-three, that he was to unify the sciences on the basis of a purely rational process. To this day, all the French are self-proclaimed Cartesians, an epithet that, in France, carries little more significance than being rational, but that nonetheless sets them apart from their neighbors!.

[3] Argand was born in Geneva on July 18, 1768.

THE NUMBER LATTICE

In the scheme we propose, the fraction y/x will be represented on the *number lattice* by a point, or *node*, whose abscissa measures x agreed-upon units of length along the x-axis and whose ordinate measures y such units along the y-axis, as shown in Figure 6.1. Each node uniquely represents a fraction, and the imaginary straight line joining that node to the origin uniquely represents a rational number. That line touches an infinity of nodes upon the grid, all of which

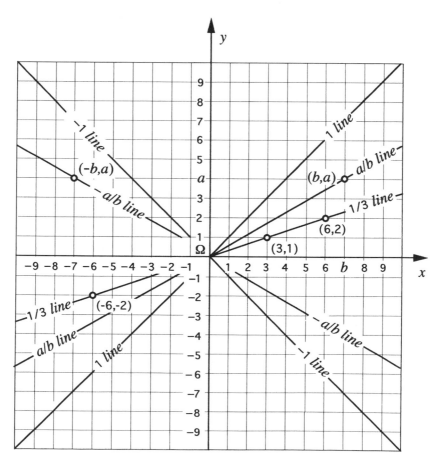

Figure 6.1. The number lattice.

correspond to equivalent fractional representations of the same rational number.

The node whose abscissa is b and whose ordinate is a will be indifferently identified as "node (b, a)" or "node a/b." The first expression places b before a for consistency with Cartesian coordinates, where x is always placed before y. The latter expression is perhaps less confusing, as it is consistent with the traditional way of representing fractions. Unless the word "node" precedes a/b, that expression must be construed as representing a fraction, not the node itself.

An infinite straight line, connecting origin $(0, 0)$ to node a/b and extending in both directions, is the locus of all integer pairs (x, y) satisfying the relationship

$$y/x = a/b, \quad \text{or} \quad ax = by.$$

On that same line, symmetrically with respect to the origin, lie those nodes (x', y') that satisfy

$$y'/x' = -a/-b = a/b.$$

That line thus uniquely represents the rational number a/b. The lattice must be understood to exist only at the nodes. A straight line is an imaginary construction that traverses the lattice, where it may or may not encounter a node upon its path. The horizontal axis is one such line, and represents the rational integer zero. Upon it may be found nodes $(x, 0)$, $(-x, 0)$. The vertical axis represents infinity, upon which may be found nodes $(0, y)$, $(0, -y)$ . . . .

Origin $(0, 0) = (\infty, \infty) = (0, \infty) = (\infty, 0)$ . . . is the point where all lines converge. It is a kind of "black hole," where all numbers merge into indeterminacy. We shall call it Ω.

In addition to the zero and infinity lines, Figure 6.1 shows line representations of the integers 1 and -1. Upon the 1 line lie those nodes (x, y) for which $y/x = 1/1 = -1/-1$. Upon the -1 line, lie those nodes (x, y) for which $y/x = -1/1 = 1/-1$.

The 0 line, the ∞ line, the 1 line, and the -1 line divide the plane into four symmetrical pairs of octants. Two symmetrical octants with respect to Ω are regarded as one. The plane thus consists of four distinct regions, each containing an infinitely dense *bundle* of lines. Since each of the four lines stands at the edge of two contiguous bundles, we may once and for all arbitrarily assign it to either of the two.

Prime Nodes

As a straight line is "shot" from origin Ω, the first node a/b that it encounters corresponds to an irreducible fraction, as integers a, b are relatively prime. Node (b, a) will be referred to as a *prime node*. Any subsequent collinear node (bk, ak) on the line's trajectory corresponds to the fraction $ak/bk = a/b$, where k is an integer. The irreducible fraction and the subsequent fractions are equivalent, meaning that they all correspond to the same rational number. Prime nodes are represented by little circles in Figure 6.2. For example, the prime nodes strung at abscissa 9 between the 0 and 1 lines represent the fractions 1/9, 2/9, 4/9, 5/9, 7/9, 8/9. They are six in number. Considering only the first octant, the number of prime nodes on any given abscissa b is Euler's function $\phi(b)$, which is defined as the number of integers that are relatively prime to integer b.

Given any two fractions, there exists an infinity of fractions that are larger than the smaller of the two, and smaller than the larger. That *everywhere-dense* distribution of rationals led the Greeks to believe that the universe Q of rational numbers was gapless, and that a one-to-one correspondence could be established between its members and the points on a line. That belief constituted the foundation

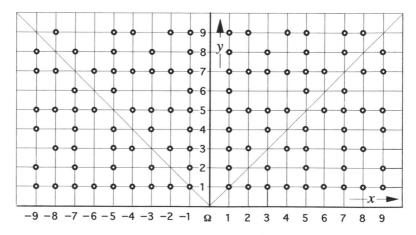

Figure 6.2. Prime nodes.

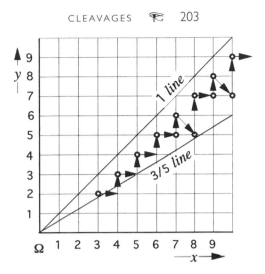

Figure 6.3. Prime node enumeration within pencil 1/1–3/5.

of their number-geometry metaphor. These prime nodes are nonetheless *denumerable*, a notion that will be discussed in the next chapter. Figure 6.3 illustrates one strategy for enumerating the prime nodes corresponding to rationals greater than 3/5 and smaller than 1. That requires testing for every value of $x = 1, 2, 3, \ldots$ those values of $y = 1, 2, 3, \ldots$ that correspond to a prime node contained within the *pencil*: 2/3, 3/4, 4/5, 5/6, 5/7, 6/7, 5/8, 7/8, 7/9, 8/9, etc.

Cleavages

If we attempt to draw a straight line representing an irrational number, that line may not touch any node, for that would mean that its slope is rational. It is thus possible, starting from origin Ω, to "shoot" hypothetical infinitely thin beams of light that travel to infinity without ever encountering a node.

What, then, is the nature of that elusive number that cannot be represented by an integer pair, and whose geometric representation cannot encounter any node upon the lattice, though these nodes are infinite in number? More specifically, how may we represent $\sqrt{2}$ on the lattice, if we cannot identify a single node, prime or not, along its path?

Imagine that the number lattice is materialized by a regularly tiled floor. Figure 6.4 represents that floor after a powerful earthquake,

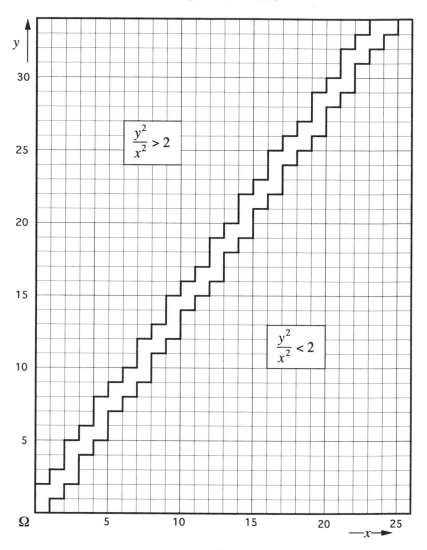

Figure 6.4. $\sqrt{2}$ Cleavage.

that cleaved its nodes into two distinct compact sets, those for which $y^2/2x^2 < 1$, and those for which $y^2/2x^2 > 1$. Each set is bounded by an irregular staircase-shaped borderline. As a matter of convention, the lower of the two will be referred to as the lower cleavage line, or simply the *cleavage line*. Because of the irrational character of $\sqrt{2}$, the light beam "shot" from origin Ω will never touch either bank of the cleavage. Upon closer examination, we will discover that the cleavage line fully defines $\sqrt{2}$, as it dramatically illustrates Dedekind's *Schnitt*.

In the left panel of Figure 6.5a, the number lattice is cleaved in such a way that those nodes for which $y/x \leq 4/3$ lie on one side, and those for which $y/x > 4/3$ lie on the other. In this case, if an infinitely thin line with a slope of $4/3$ is drawn from origin Ω, it touches every third node of the lower set, and flies above the other two. In so doing, it never touches a node belonging to the upper set. That cleavage represents Dedekind's *Schnitt* for a rational number, where it was decided that the rational number itself belongs to the lower of the two sets, in other words, that the lower set is top-bound.

The right panel of Figure 6.5a shows a cleavage of the number lattice such that those nodes for which $y/x \geq 4/3$ lie on one side, and those for which $y/x < 4/3$ lie on the other. In this case, if an infinitely thin line with slope $4/3$ is drawn from origin Ω, it touches every third node of the upper set, and flies below the other two. In so doing, it never touches a node belonging to the lower set. That cleavage represents Dedekind's *Schnitt* for a rational number, where it was decided that the rational number itself belongs to the upper set, which is bottom-bound in this case. The top-bound cleavage option (left panel of Figure 6.5a) can be made to match the bottom-bound option (right panel), by turning either diagram upside down and sliding it against the other until coincidence is achieved.

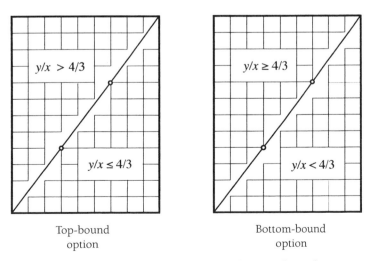

Top-bound
option

Bottom-bound
option

Figure 6.5a. 4/3 cleavage, top-bound and bottom-bound options.

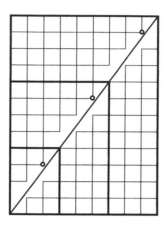

Figure 6.5b. Nodes $(3, 2)$, $(7, 5)$, $(11, 8)$.

In what follows, we shall adopt the top-bound option once and for all, and limit our discusion to positive real numbers, as that discussion may be easily extended to negative numbers as well.

With that in mind, consider the bottom left rectangle of Figure 6.5b, whose width is 3 and height 4. It is diagonally traversed by the line representing the rational number 4/3. The smallest rational upon the upper cleavage line is $(2, 3)$. If the rectangle under consideration is doubled in size, that number is $(5, 7)$. Generally, the smallest rational number larger than 4/3 in a rectangle of width $3m$ and height $4m$, where m is an integer, is $(4m - 1)/(3m - 1)$. Its successive values are 3/2, 7/5, 11/8, 15/11 ... 399/299, ... That fraction gets smaller and smaller with increasing values of m, suggesting that there exists no *smallest rational* larger than 4/3 above the diagonal.

The cleavage metaphor thus illustrates the following important aspect of the Dedekind *Schnitt*: when the lower set is top-bound, there exists no smallest rational in the upper set, and conversely, when the upper set is bottom-bound, there exists no greatest rational in the lower set. In an irrational cleavage, there exists no greatest rational in the lower set, and no smallest rational in the upper set.

In what follows, $\partial_\mu(x)$ will denote the magnitude of the cleavage line's *increment*, or upward jump at abscissa x, corresponding to the number μ. The term $h_\mu(x)$ will denote that line's *height* at abscissa x. The term Δ_μ will represent the infinite increment sequence. The term H_μ will denote the infinite cleavage height sequence for consecutive values of x, and will be called the *cleavage sequence* for the number μ.

If $[N]$ denotes the integral part of the number N, it is clear that for any number μ,

$$h_\mu(x) = [x\mu], \qquad (6.1)$$

or

$$h_\mu(x) \leq \mu x < h_\mu(x) + 1, \qquad (6.2)$$

where $h_\mu(x)$ is an integer. Statements (6.1) and (6.2) are equivalent definitions for $h_\mu(x)$.

Calculating $h_\mu(x)$ for a given value of x is done by assigning the values $0, 1, 2, \ldots$ to y, and for each value, performing the test: Is $y > \mu x$? The highest value of y for which the answer is negative is $h_\mu(x)$.

For the rational number $\mu = a/b$, the test is, Is $by > ax$? (6.3)

For $\mu = \sqrt{2}$, the test is, Is $y^2 > 2x^2$? (6.4)

For $\mu = (a + \sqrt{b})/c$, the test is, Is $(cy - ax)^2 > bx^2$? (6.5)

The process may, of course, be stretched to any desired length. The algorithm is workable only inasmuch as the test may be realized, which is not always obvious in the absence of statements as simple as $(\sqrt{2})^2 = 2$. That is the case with the transcendental numbers e and π, among others. Tests may be imagined, however, based on known developments of numbers other than their positional representation, such as

$$e = 1 + \frac{1}{1!} + \frac{1}{2!} + \frac{1}{3!} + \cdots, \qquad (6.6)$$

from which the following specific test may be derived:[4]

$$\text{Is } y > x\left(1 + \frac{1}{1!} + \frac{1}{2!} + \frac{1}{3!} + \cdots + \frac{1}{x!}\right) ? \qquad (6.7)$$

---

[4]The proof of that statement is given in the appendix to this chapter.

The rational number $\theta_\mu(x) = h_\mu(x)/x$ will be referred to as the *lower slope of node* $(x, h_\mu(x))$, or simply the *slope of* μ *at* x. Similarly, $\Theta_\mu(x) = (h_\mu(x) + 1)/x$ will be referred to as the *upper slope of that node*, or the *upper slope of* μ *at* x. From the previous definitions, $\theta_\mu(x) \le \mu$ and $\Theta_\mu(x) > \mu$ for all values of x. The integer pair $(x, h_\mu(x))$ is said to *belong* to μ. Clearly, as x tends to infinity, $h_\mu(x) + 1 \to h_\mu(x)$, and both the upper and lower slopes converge to μ.

The increment sequence $\Delta_{4/3} = 1\,1\,2,\ 1\,1\,2,\ 1\,1\,2\ldots$ is *periodic*, and its period equals 3, the denominator of irreducible fraction 4/3. The sum of any three consecutive increments is equal to the fraction's numerator, 4. The cleavage sequence $H_{4/3} = 1\,2\,4\,5\,6\,8\,9\,10\,12\ldots$ is *regular*. When it is shifted by three places or multiples thereof to the left or right and compared term for term with the original, the difference between corresponding terms is constant. Any segment of the cleavage line of width 3 and height 4 repeats itself indefinitely, as it slides upon its diagonal. The cleavage sequence is regular, and the increment sequence periodic, if and only if μ is rational.

Tables 6.1 and 6.2 show the increment sequences for $\mu = \sqrt{2}$ and $\mu = \sqrt{3}$, respectively. Their increment sequences both consist of a succession of 1s and 2s, and only the pattern differs. The sequences are to be read as one reads a paragraph of text. Sequence $\Delta_{\sqrt{2}}$ consists of packets of single and double ones, separated by a single two, and

TABLE 6.1

Increment sequence for $\sqrt{2}$.

```
1 1 2 1 2 1 1 2 1 2 1 1 2 1 2 1 1 2 1 2 1 1
2 1 2 1 2 1 1 2 1 2 1 1 2 1 2 1 2 1 2 1 1
2 1 2 1 1 2 1 2 1 2 1 1 2 1 2 1 1 2 1 2 1 1 2 1
2 1 2 1 1 2 1 2 1 2 1 1 2 1 2 1 2 1 1 2 1 1 2 1
2 1 2 1 1 2 1 2 1 1 2 1 2 1 1 2 1 2 1 2 1 1 2 1
2 1 1 2 1 2 1 2 1 1 2 1 2 1 1 2 1 2 1 1 2 1 2 1
2 1 1 2 1 2 1 1 2 1 2 1 2 1 1 2 1 2 1 1 2 1 2 1
2 1 1 2 1 2 1 1 2 1 2 1 1 2 1 2 1 1 2 1 2 1
1 2 1 2 1 2 1 1 2 1 2 1 1 2 1 2 1 2 1 2 1
1 2 1 2 1 1 2 1 2 1 2 1 1 2 1 1 2 1 2 1 1 2
1 2 1 2 1 1 2 1 2 1 1 2 1 2 1 2 1 1 2 1 1 2
1 2 1 2 1 1 2 1 2 1 1 2 1 2 1 2 1 2 1 1 2
1 2 1 1 2 1 2 1 2 1 1 2 1 2 1 1 2 1 1 2 1 2
1 2 1 1 2 1 2 1 1 2 1 2 1 2 1 1 2 1 1 2 1 2
1 2 1 1 2 1 2 1 1 2 1 2 1 1 2 1 2 1 1 2 1 2
1 1 2 1 2 1 1 2 1 2 1 1 2 1 2 1 2 1 2 1 2
1 1 2 1 2 1 1 2 1 2 1 2 1 1 2 1 2 1 1 2 1 1
2 1 2 1 2 1 1 2 1 2 1 1 2 1 2 1 2 1 2 1 1
2 1 2 1 2 1 1 2 1 2 1 1 2 1 2 1 2 1 2 1 1
2 1 2 1 1 2 1 2 1 2 1 1 2 1 2 1 1 2 1 2 1 1
```

TABLE 6.2

Increment sequence for $\sqrt{3}$.

1	2	2	1	2	2	2	1	2	2	2	1	2	2	1	2	2	2	1	2	2	2	1	2	2	2	
1	2	2	1	2	2	2	1	2	2	2	1	2	2	2	1	2	2	1	2	2	2	1	2	2	2	
1	2	2	1	2	2	2	1	2	2	2	1	2	2	2	1	2	2	1	2	2	2	1	2	2	2	
1	2	2	2	1	2	2	1	2	2	2	1	2	2	2	1	2	2	2	1	2	2	1	2	2	2	
1	2	2	2	1	2	2	1	2	2	2	1	2	2	2	1	2	2	2	1	2	2	1	2	2	2	
1	2	2	2	1	2	2	2	1	2	2	1	2	2	2	1	2	2	2	1	2	2	2	1	2	2	
1	2	2	2	1	2	2	2	1	2	2	1	2	2	1	2	2	2	1	2	2	2	1	2	2		
1	2	2	2	1	2	2	2	1	2	2	2	1	2	2	1	2	2	2	1	2	2	2	1	2	2	
1	2	2	2	1	2	2	2	1	2	2	2	1	2	2	1	2	2	2	1	2	2	2	1	2	2	
2	1	2	2	1	2	2	2	1	2	2	2	1	2	2	2	1	2	2	1	2	2	2	1	2	2	
2	1	2	2	1	2	2	2	1	2	2	2	1	2	2	2	1	2	2	1	2	2	2	1	2	2	
2	1	2	2	2	1	2	2	1	2	2	2	1	2	2	2	1	2	2	2	1	2	2	1	2	2	
2	1	2	2	2	1	2	2	1	2	2	2	1	2	2	2	1	2	2	2	1	2	2	1	2	2	
2	1	2	2	2	1	2	2	2	1	2	2	1	2	2	2	1	2	2	2	1	2	2	2	1	2	
2	1	2	2	2	1	2	2	2	1	2	2	1	2	2	2	1	2	2	2	1	2	2	2	1	2	
2	1	2	2	2	1	2	2	2	1	2	2	2	1	2	2	1	2	2	2	1	2	2	2	1	2	
2	1	2	2	2	1	2	2	2	1	2	2	2	1	2	2	1	2	2	2	1	2	2	2	1	2	
2	2	1	2	2	1	2	2	2	1	2	2	2	1	2	2	2	1	2	2	1	2	2	2	1	2	
2	2	1	2	2	1	2	2	2	1	2	2	2	1	2	2	2	1	2	2	1	2	2	2	1	2	
2	2	1	2	2	2	1	2	2	1	2	2	2	1	2	2	2	1	2	2	2	1	2	2	1	2	

sequence $\Delta_{\sqrt{3}}$ consists of packets of double and triple twos, separated by a single one. The sequence arrangement enhances the nonperiodic character of the sequences, while suggesting their pattern. These patterns are analyzed in Appendix 6.2.

Whether μ is rational or irrational, the cleavage sequence is *monotone*, meaning that its successive terms may not decrease as x increases. Clearly, $\mu > \mu'$ if and only if there exists a value of x such that $h_\mu(x) > h_{\mu'}(x)$. As x increases, the corresponding cleavage lines drift further and further apart. They may merge again and separate again, but eventually they separate once and for all. In other words, given the numbers $\mu > \mu'$, and the arbitrary integer d, a value x_d of x may always be found beyond which $h_{\mu(x)} - h_{\mu'}(x) > d$ for all $x > x_d$.

Consider the example of Figure 6.6, where $\mu = \sqrt{2}$ and $\mu' = 15/11$ (note that node $(11, 15)$ belongs to $\sqrt{2}$). The cleavage lines coincide until $x = 4$. With the next value, namely, $x = 5$, which we may call the *initial separation abscissa* x_s, we get

$$7/5 < \sqrt{2} < 8/5 : \qquad h_{\sqrt{2}}(5) = 7,$$

while

$$6/5 < 15/11 < 7/5 : \quad h_{15/11}(5) = 6.$$

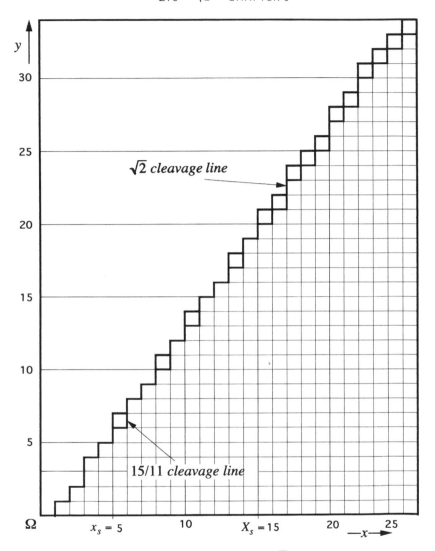

Figure 6.6. Cleavage lines for $\sqrt{2}$ and 15/11.

The cleavage lines again coincide for $x = 6, 7, 9, 11, 12, 14$, beyond which $h_{\sqrt{2}}(x) > h_{15/11}(x)$ for all subsequent values of x, as the cleavage lines slowly drift away from one another. Abscissa 15 will be referred to as the *final separation abscissa* and denoted X_s.

Generally, for any given $\mu' < \mu$, there exist integers x_s and X_s such that

$$h'_\mu(x) = h_\mu(x) \quad \text{for } x < x_s, \tag{6.8a}$$

$$h_{\mu'}(x) \leq h_\mu(x) \quad \text{for } x_s \leq x < X_s, \tag{6.8b}$$

$$h_{\mu'}(x) < h_\mu(x) \quad \text{for } x \geq X_s. \tag{6.8c}$$

To further illustrate that point, Table 6.3 shows the values of x_s and X_s for $\mu = \sqrt{2}$ and various values of μ' that consist of increasingly longer chunks of the decimal representation of $\sqrt{2}$, all of which belong to $\sqrt{2}$. The table suggests that a cleavage sequence may not uniquely

TABLE 6.3
Initial and final separation abscissas for $\mu = \sqrt{2}$ and a selection of values μ'.

	μ'				
	1	1.4	1.41	1.414	1.4142
x_s	3	17	17	99	169
$h_\mu(x_s)$	4	24	24	140	239
X_s	4	72	224	4589	73312
$h_\mu(X_s)$	5	100	315	6489	103678

define a number unless it is allowed to extend to infinity. To any number corresponds one and only one infinite increment sequence as well as one and only one cleavage sequence, and conversely. We shall say that H_μ defines μ.

Coherence

Nodes (b, a) and (d, c) are said to be *coherent* if a number exists to which they both belong. For instance, nodes $(5, 7)$ and $(17, 24)$, both of which are known to belong to $\sqrt{2}$, are coherent. So are nodes $(5, 7)$ and $(17, 23)$, both of which belong to 1.41. On the other hand, there is no number to which $(5, 7)$ and $(17, 22)$ both belong, as whatever number the latter node belongs to must be smaller than $23/17$, whereas $7/5 > 23/17$. It is easily shown that the necessary and sufficient condition for nodes (b, a) and (d, c) to be coherent is

$$\frac{a+1}{b} > \frac{c}{d} \quad \text{and} \quad \frac{c+1}{d} > \frac{a}{b}. \tag{6.9a}$$

The "pencil" of lines whose slope is equal to or greater than a/b and smaller than $(a + 1)/b$ will be denoted $P(b, a)$. The geometric interpretation of statement (6.9a) is that in order for nodes (b, a) and (d, c) to be coherent, it is necessary and sufficient that the pencils $P(b, a)$ and $P(d, c)$ overlap.

Any number that belongs to both nodes must be equal to or greater than the greater of a/b and c/d and smaller than the smaller of $(a + 1)/b$ and $(c + 1)/d$. For example, coherent nodes $(5, 7)$ and $(8, 11)$ both belong to any number μ such that

$$\frac{7}{5} \leq \mu < \frac{12}{8}.$$

The range of μ in the above statement is precisely the extent of the overlap of $P(5, 7)$ and $P(8, 11)$, as shown in Figure 6.7. One eligible number μ is $\sqrt{2}$, whose cleavage line is also shown.

If nodes (b, a) and (d, c) are coherent, and so are nodes (d, c) and (f, e), that does not necessarily entail that (b, a) and (f, e) are coherent. In other words, the property of coherence, though symmetric and reflexive, is not transitive. For example $(16, 22)$ and $(17, 24)$ are coherent, and so are $(16, 22)$ and $(17, 23)$, whereas nodes $(17, 23)$ and $(17, 24)$ are obviously not coherent, as they correspond to two distinct cleavage heights for the same abscissa.

Let S be a set containing a *finite* collection of nodes. The necessary and sufficient condition for a number μ to exist to which all of the nodes belong is that they be *coherent in pairs*, in other words, that each node be coherent with every other node in the set. Any such number μ is equal to or greater than the greatest lower slope and smaller than the smallest upper slope in S. A set all of whose nodes are coherent in pairs is said to be coherent. The set S is said to be coherent with, or belong to, μ.

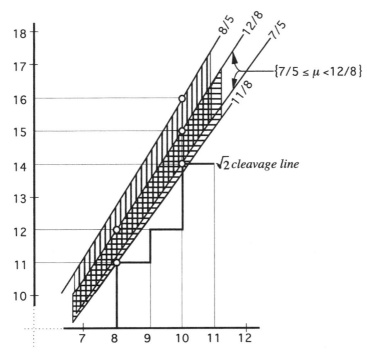

Figure 6.7. Overlap of pencils $P(5, 7)$ and $P(8, 11)$ showing the $\sqrt{2}$ cleavage line.

To any finite coherent node set S thus corresponds an infinity of numbers to which it belongs. For example,

$$S = \{(1, 1), (4, 5), (7, 9), (9, 12), (11, 15), (14, 19)\}$$

is coherent with both $\sqrt{2}$ and 15/11, amongst an infinity of numbers. Adding node $(8, 11)$ to the set preserves its coherence with $\sqrt{2}$, but not with 15/11. Similarly, adding node $(13, 17)$ to the set preserves its coherence with 15/11, but not with $\sqrt{2}$. On the other hand, the sets

$$S = \{(1, 1), (4, 5), (7, 9), (9, 12), (11, 15), (14, 19)\}$$

and

$$S' = \{(3, 4), (8, 11), (15, 21), (18, 25)\}$$

are both coherent with $\sqrt{2}$, though disjoined (nonoverlapping).

If nodes a/b and c/d are such that

$$\frac{a}{b} \geq \frac{c}{d} \quad \text{and} \quad \frac{a+1}{b} < \frac{c+1}{d} \tag{6.9b}$$

node a/b is said to *subsume* node c/d. The two nodes are obviously coherent, but additionally the pencil $P(b, a)$ is entirely contained with the pencil $P(d, c)$. The reader may easily show that inequalities in (6.9b) imply that $b > d$. Consider, for example, nodes 24/17 and 4/3, both of which belong to $\sqrt{2}$. Obviously, 24/17 > 4/3. On the other hand, 25/17 < 5/3. Node 24/17 therefore subsumes node 4/3. In the next example, $a/b = c/d$. (Note that the nodes under consideration need not necessarily be prime.) In this example, we shall pick nodes 4/3 and 16/12, both of which also belong to $\sqrt{2}$. We get 17/12 < 5/3, meaning that node 16/12 subsumes node 4/3, a conclusion that should have been obvious from the outset.

When enumerating a set of coherent nodes in order of ascending values of the abscissa, if we come upon a node that subsumes a previous node in the enumeration, the latter node becomes redundant. In other words, it does not contribute anything new to out attempt to zero in on whatever number the coherent node set belongs to. Let us return to set $S = \{(1, 1), (4, 5), (7, 9), (9, 12), (11, 15), (14, 19)\}$, which is coherent with both $\sqrt{2}$ and 15/11. If we add node 239/169, the set's coherence with $\sqrt{2}$ is preserved, and actually enhanced, as node 239/269 subsumes every preceding node in the set. Set S might just as well consist of that single node.

Let M represent the set of *all* nodes belonging to μ. Its elements are those of sequence H_μ, irrespective of the sequence's order. M is infinite and denumerable (a notion that will be clarified in the next chapter). An infinite number of proper subsets of M may be created, all of which are infinite, denumerable, and coherent with μ. The number μ may thus be coherent with an infinity of infinite node sets, all of which are proper subsets of M.

If $\mu' \neq \mu$, there exists a value X_s such that $h_{\mu'}(x) \neq h_\mu(x)$ for all $x \geq X_s$. Though finite subsets of M may be found that are coherent with μ' when all of their nodes have abscissas smaller than X_s, no subset of M, finite or infinite, may be coherent with μ' if it contains a node whose abscissa is equal to or greater than X_s. Consequently, no *infinite* subset of M may be found that is coherent with any number other than μ. Any infinite node set that is coherent with a given number thus uniquely defines that number (defines that number and only that number). That infinite set does not necessarily contain every node in H_μ. It may have infinitely many gaps between successive nodes, and each gap may itself contain an infinite number of nodes.

A Definition of Real Numbers

It is believed that Thales of Miletus, the father of Greek geometry, indeed, of Greek mathematical thought, was the first to grasp the importance of the notion of geometrical locus, that curve whose points all share a particular property in terms of the framework within which the curve is drawn. Centuries later, the discovery of conics, following the intense and fruitless search for the solution of the three famous problems of antiquity, namely, the squaring of the circle, the doubling of the cube, and the trisection of an angle, led the Greeks to sharpen their understanding of geometric loci. The ellipse was thus defined as the locus of all points the sum of whose distances from two fixed points is constant, and the great Apollonius defined the parabola as the curve "whose semichord is the mean proportional between its Latus rectum and the height of the segment." The above rhetorical definitions were esentially intrinsic, in that the curves were defined in terms of elements thereof. The momentous contribution of Descartes, and to a certain extent of Fermat, was to introduce a universal *referential*, namely, the orthogonal coordinate system, in terms of which algebraic expressions could be given for curves. The definition of Apollonius could thus be rephrased as $x^2 = Ly$, where abscissa x is the the semichord, ordinate y is the segment's height, and L is the latus rectum. Nonetheless, that *extrinsic* definition is itself tantamount to that of a locus, namely, that of all pairs of abscissas x and ordinates y such that the square of the first is proportional to the second, where L is the proportionality coefficient.

Moving on to the domain of number, and borrowing from the language of geometry, we may state that the rational number ρ is the *locus* of those integer pairs (x, y), which are infinite in number, sharing the property $y = \rho x$. The geometric equivalent of that locus on the number lattice is the straight line connecting all nodes (x, y) satisfying $y = \rho x$. That rather coarse definition of a rational number in terms of an infinite set of integer pairs was only intended to introduce the reader to the notion of locus in the realm of number. That definition may now be refined, and extended to the domain of (positive) real numbers:

Real number μ is the locus of those integer pairs (x, y) for which $y \leq \mu x < y + 1$, over any infinite set of integral values of x.

It follows from the above discussion that the number lattice need not include origin Ω (or the y-axis itself, for that matter), as long as it is infinite in the x and y directions. One may thus partly escape the conceptual difficulties raised by the ubiquitous nature of Ω, unfortunately without escaping those inherent in the nature of infinity itself. That statement is consistent with the stipulation that the Dedekind *Schnitt* traverse an infinite set of rationals, and is reminiscent of the positional number notation: whether a number is rational or irrational, its representation is essentially infinite.

Consider an "immensely large" finite set S that is coherent with, say, $\sqrt{2}$. We may add to it a new node that subsumes every node in S. The new set may thus be reduced to that node alone. We now add to this set a new node that subsumes its only member, and takes its place. That process may be continued indefinitely, as there is no "largest number." Had there been such a thing, as some ancient Greeks believed, any number could have been represented by that mythical ultimate node. Unfortunately, things are not that simple, and our only recourse is to resort to an infinite cloud of nodes that brings us as close to the desired number as we wish. This little exercise prepares the reader for the next and final chapter, which deals with infinity.

SOME PROPERTIES OF FRACTIONS

Consider the fractions

$$\frac{a}{b} > \frac{c}{d}. \tag{6.10}$$

That inequality obviously implies that

$$ad - bc > 0. \tag{6.11}$$

What are the necessary and sufficient conditions for the fraction y/x to satisfy

$$\frac{a}{b} > \frac{y}{x} > \frac{c}{d} \ ? \tag{6.12}$$

In other words, that

$$ax - by > 0 \quad \text{and} \quad dy - cx > 0 \; ? \tag{6.13}$$

Let us pick two integers $k, l \geq 1$ at random, and write

$$x = kb + ld, \qquad y = ka + lc, \tag{6.14}$$

yielding

$$\frac{y}{x} = \frac{ka + lc}{kb + ld}. \tag{6.15}$$

It is easily shown that the values thus obtained for x, y constitute a set of sufficient conditions for (6.12) to be verified. Indeed,

$$ax - by = a(kb + ld) - b(ka + lc) = l(ad - bc), \tag{6.16a}$$

and

$$dy - cx = d(ka + lc) - c(kb + ld) = k(ad - bc). \tag{6.16b}$$

Given (6.11), equations (6.16a) and (6.16b) result in $ax - by > 0$ and $dy - cx > 0$. Hence,

$$\frac{a}{b} > \frac{ka + lc}{kb + ld} > \frac{c}{d}. \tag{6.17}$$

It is equally easy to show that any fraction y/x that satisfies (6.12) must necessarily be of the same *form* as (6.17). Assuming (6.12) to be true, we may write

$$ax - by = n \quad \text{and} \quad dy - cx = m, \tag{6.18}$$

where m, n are integers ≥ 1. These two equations together yield

$$x(ad - bc) = mb + nd \quad \text{and} \quad y(ad - bc) = ma + nc.$$

We shall call the positive integer $ad - bc = D$ the *determinant* of a/b and c/d. We obtain from (6.18)

$$Dy = ma + nc \quad \text{and} \quad Dx = mb + nd, \tag{6.19}$$

which yields

$$\frac{y}{x} = \frac{ma + nc}{mb + nd}, \qquad \text{where } m, n \geq 1. \qquad (6.20)$$

Statement (6.17) signifies that given two fractions $a/b > c/d$, the fraction y/x such that $x = kb + ld$, and $y = ka + lc$, where k, l are any integers ≥ 1, satisfies $a/b > y/x > c/d$. Conversely, given three fractions $a/b > y/x > c/d$, while it may not always be possible to find two integers $k, l \geq 1$ such that $x = kb + ld$ and $y = ka + lc$, statement (6.19) signifies that it is always possible to find two integers $m, n \geq 1$ such that $Dy = ma + nc$ and $Dx = mb + nd$, in other words, such that $\dfrac{y}{x} = \dfrac{ma + nc}{mb + nd}$.

Example 6.1

Let $\dfrac{a}{b} = \dfrac{11}{13}, \dfrac{c}{d} = \dfrac{5}{7}$ and let $k = 3, l = 5$.
We get $x = 39 + 35 = 74$, $y = 33 + 25 = 58$. Hence, $y/x = 58/74 = 29/37$, and we verify that $11/13 > 29/37 > 5/7$.

Example 6.2

Let $\dfrac{a}{b} = \dfrac{11}{13}, \dfrac{c}{d} = \dfrac{5}{7}$, and $\dfrac{y}{x} = \dfrac{37}{47}$.
We verify that $\dfrac{11}{13} > \dfrac{37}{47} > \dfrac{5}{7}$, and we get

$$m = dy - cx = 24, \quad n = ax - by = 36, \quad \text{and } D = ad - bc = 12.$$

In this particular case, both m and n are divisible by D. This allows us to write $k = m/D = 2$ and $l = n/D = 3$, and verify that

$$x = 2b + 3d = 26 + 21 = 47 \quad \text{and} \quad y = 2a + 3c = 22 + 15 = 37.$$

Example 6.3

Let $\dfrac{a}{b} = \dfrac{10}{13}, \dfrac{c}{d} = \dfrac{8}{11}$, and $\dfrac{y}{x} = \dfrac{59}{81}$.

We verify that $\dfrac{10}{13} > \dfrac{59}{81} > \dfrac{8}{11}$, and we get

$$m = dy - cx = 1, \qquad n = ax - by = 43, \qquad \text{and} \quad D = ad - bc = 6,$$

which yields

$$x = \frac{b + 43d}{6} = \frac{486}{6} = 81 \quad \text{and} \quad y = \frac{a + 43c}{6} = \frac{354}{6} = 59.$$

Note that in this case neither m nor n is divisible by D. However, both $ma + nc$ and $mb + nd$ are divisible by D.

Why did we choose to call D the determinant? To answer that question, let us turn to the two equations comprising (6.18) and express them in matrix form. We get

$$\begin{bmatrix} m \\ n \end{bmatrix} = \begin{bmatrix} d & -c \\ -b & a \end{bmatrix} \times \begin{bmatrix} y \\ x \end{bmatrix}. \tag{6.21a}$$

The determinant of that matrix is none other than $ad - bc = D$. From (6.11), that determinant is nonzero, meaning that the matrix is regular, or that it possesses one and only one inverse matrix. The reader who is even slightly familiar with matrix algebra will be able to conclude from (6.21a) that

$$\begin{bmatrix} y \\ x \end{bmatrix} = \frac{1}{D} \begin{bmatrix} a & c \\ b & d \end{bmatrix} \times \begin{bmatrix} m \\ n \end{bmatrix}, \tag{6.21b}$$

which statement is identical to (6.19).

Contiguous Fractions

We shall say of two fractions the absolute value of whose determinant D is equal to 1 that they are *contiguous*. We could have coined a name other than "contiguous" such as "unilogous," which makes reference to the determinant's value of one. For reasons that will become apparent later, we could have also used the name "Stern-Brocot fractions," and "affine fractions." The larger of the two will be referred to as the *predecessor*, and the other as the *successor*. In the preceding examples, fractions 59/81 and 8/11 are contiguous. Indeed,

$59 \times 11 - 8 \times 81 = 649 - 648 = 1$. Fraction 59/81 is the predecessor, and 8/11 the successor.

Let us return to fractions $a/b > c/d$, and assume that

$$ad - bc = 1. \tag{6.22a}$$

This will be written

$$\frac{a}{b} \Rightarrow \frac{c}{d}. \tag{6.22b}$$

Clearly any two integers A, B, the absolute value of whose difference is equal to 1, cannot share a common divisor greater than 1. Thus 1 is their greatest common divisor (GCD), and the statement is written $GCD(A, B) = 1$, or sometimes simply $(A, B) = 1$. A and B are said to be *relatively prime*, a relationship that is also written $A \perp B$. Turning to (6.22a), it follows that $bc \perp ad$, which entails the relationships

$$a \perp b, \ c \perp d, \ a \perp c, \ b \perp d. \tag{6.22c}$$

The first two relationships imply that no fraction may belong to a contiguous pair unless it is reduced to its lowest terms. The last two relationships imply that no two fractions may be contiguous unless their numerators are relatively prime, and their denominators as well. For example,

$$\frac{4}{5} \Rightarrow \frac{11}{14} \quad \text{and} \quad 4 \perp 5, \ 11 \perp 14, \ 4 \perp 11, \ 5 \perp 14.$$

Observe that (6. 22a) also signifies that $a/c \Rightarrow b/d$.

We now ask ourselves the following question: given any relatively prime pair (a, b), is it always possible to find some pair (y, x) such that $a/b \Rightarrow y/x$? One of the most elementary indeterminate Diophantine problems consists, given positive integers a, b, of finding positive integers x, y satisfying

$$ay - bx = c.$$

It is possible to show that solutions exist if c is a multiple of $GCD(a, b)$. When $a \perp b$, the equation $ay - bx = 1$ satisfies that condition, and can therefore be solved for x, y. For example, $3/5 \Rightarrow 4/7, 7/12, 10/17, \ldots 121/202, \ldots$.

It is easy to also show that any irreducible fraction is the successor of an infinity of contiguous fractions.

Equation (6.22a) also implies that

$$a, d \geq 1 \quad \text{and} \quad b, c \geq 0. \tag{6.22d}$$

The contiguous pair corresponding to $a, d = 1$, and $b, c = 0$ is

$$\frac{1}{0} \Rightarrow \frac{0}{1}, \tag{6.22e}$$

where it is posited that $0 \perp 1$.

As a further example, we can define the Fibonacci numbers[5] of order m as follows:

$$F_{m, n+2} = F_{m, n} + mF_{m, n+1}, \quad \text{with } F_{m, 0} = F_{m, 1} = 1. \tag{6.23a}$$

For $m = 1$, we obtain the classical Fibonacci sequence

$$0, 1, 1, 2, 3, 5, 8, 13, 21, \ldots, \tag{6.23b}$$

which yields the sequence of ratios $F_{m, n+1}/F_{m, n}$ for $n = 0, 1, 2, \ldots$, with the following interesting property:

$$\frac{1}{0} \Rightarrow \frac{1}{1} \Leftarrow \frac{2}{1} \Rightarrow \frac{3}{2} \Leftarrow \frac{5}{3} \Rightarrow \frac{8}{5} \Leftarrow \cdots \tag{6.23c}$$

With the exception of the first, these ratios are the convergents of the regular periodic continued fraction $[1, 1, 1, 1, \ldots]$.

With $m = 2$, we get

$$\frac{1}{0} \Rightarrow \frac{2}{1} \Leftarrow \frac{5}{2} \Rightarrow \frac{12}{5} \Leftarrow \frac{29}{12} \Rightarrow \frac{70}{29} \Leftarrow \cdots \tag{6.23d}$$

Statements (6.23c) and (6.23d) are none other than restatements of Huygens's theorem (Equation 5.13).

---

[5] Midhat J. Gazalé, *Gnomon: From Pharaohs to Fractals* (Princeton, N.J.: Princeton University Press, 1999).

The Mediant

Let us now turn our attention to the fraction $\dfrac{y}{x} = \dfrac{a+c}{b+d}$, which is called the *mediant* of a/b and c/d. We get

$$dy - cx = d(a+c) - c(b+d) = ad - bc = 1$$

and

$$ax - by = a(b+d) - b(a+c) = ad - bc = 1.$$

In other words

$$\frac{a}{b} \Rightarrow \frac{a+c}{b+d} \Rightarrow \frac{c}{d}. \tag{6.24a}$$

Obviously, the fraction $(a+c)/(b+d)$ is itself irreducible. Be careful not to assume that the above relationship is transitive, despite appearances. Indeed, it follows from (6.24a) that

$$\frac{a}{b} \Rightarrow \frac{2a+c}{2b+d} \Rightarrow \frac{a+c}{b+d} \Rightarrow \frac{a+2c}{b+2d} \Rightarrow \frac{c}{d}, \tag{6.24b}$$

which does not entail, for instance, that $a/b \Rightarrow (a+2c)/(b+2d)$, since

$$a(b+2d) - b(a+2c) = 2(ad - bc) = 2.$$

The above procedure may be indefinitely continued, yielding

$$\frac{a}{b} \Rightarrow \frac{3a+c}{3b+d} \Rightarrow \frac{2a+c}{2b+d} \Rightarrow \frac{3a+2c}{3b+2d} \Rightarrow \frac{a+c}{b+d} \Rightarrow \frac{2a+3c}{2b+3d}$$
$$\Rightarrow \frac{a+2c}{b+2d} \Rightarrow \frac{a+3c}{b+3d} \Rightarrow \frac{c}{d}. \tag{6.24c}$$

and so on.

Generally, given $a/b \Rightarrow c/d$, assume that the relationship

$$\frac{ka + lc}{kb + ld} \Rightarrow \frac{y}{x} \tag{6.25a}$$

is verified for a given quadruple of integers k, l, x, y. By virtue of matrix equations (6.21a) and (6.21b), there exists one and only one pair of integers m, n satisfying $ma + nc = y$ and $mb + nd = x$, where the determinant D is equal to 1. We may therefore write

$$\frac{ka + lc}{kb + ld} \Rightarrow \frac{ma + nc}{mb + nd}. \tag{6.25b}$$

Hence,

$$kmab + knad + lmbc + lncd$$

$$-kmab - knbc - lmad - lncd = 1,$$

i.e. $$knad + lmbc - knbc - lmad = 1, \tag{6.25c}$$

or $$(kn - lm)(ad - bc) = 1,$$

and since $(ad - bc) = 1$, Equation (6.25b) is verified when $(kn - lm) = 1$. In other words when

$$k/l \Rightarrow m/n, \quad \text{or} \quad k/m \Rightarrow l/n. \tag{6.26}$$

This implies in turn that $k, n \geq 1$ and $l, m \geq 0$, as well as $k \perp l, m \perp n, k \perp m, l \perp n$.

Example 6.4

Let us put $a = 59$, $b = 81$, $c = 8$, $d = 11$; these values verify

$$\frac{59}{81} \Rightarrow \frac{8}{11}.$$

Let us also put $k = 4$, $m = 7$, $l = 1$, $n = 2$; these values verify

$$\frac{4}{7} \Rightarrow \frac{1}{2}.$$

We get

$$\frac{ka + lc}{kb + ld} = \frac{4 \times 59 + 1 \times 8}{4 \times 81 + 1 \times 11} = \frac{244}{335},$$

Similarly,

$$\frac{ma + nc}{mb + nd} = \frac{7 \times 59 + 2 \times 8}{7 \times 81 + 2 \times 11} = \frac{429}{589},$$

and we verify that

$$\frac{244}{335} \Rightarrow \frac{429}{589}.$$

Note that the leftmost relationship in (6.24c) namely $a/b \Rightarrow (3a + c)/(3b + d)$, verifies (6.25b), with $k = 1$, $l = 0$, $m = 3$, $n = 1$, which is perfectly legitimate, since $1/3 \Rightarrow 0/1$. Similarly, the rightmost relationship in (6.24c) namely, $(a + 3c)/(b + 3d) \Rightarrow c/d$, is also legitimate, as $k = 1$, $l = 3$, $m = 0$, $n = 1$, with $1/0 \Rightarrow 3/1$. Note also that the relationship $a/b \Rightarrow c/d$ itself corresponds to $k = n = 1$ and $l = m = 0$.

Affine Transformations

Statement (6.26) implies that when $a/b \Rightarrow c/d$, the fraction $y = (ma + nc)$, $x = (mb + nd)$ may belong to a contiguous pair, and therefore that it is irreducible, if and only if m and n are relatively prime. To every relatively prime pair m, n thus corresponds one and only one pair of relatively prime integers $y = mb + nd$ and $x = ma + nc$, and conversely. Consider, for example, the two contiguous fractions $a/b = 1/1$ and $c/d = 1/2$.

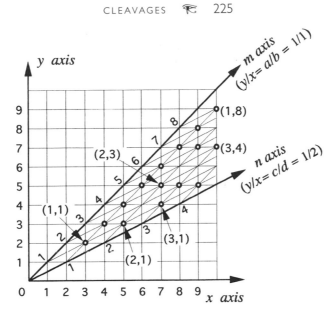

Figure 6.8a. Nodes $y/x = (ma + nc)/(mb + nd)$ for $a/b = 1/1$, $c/d = 1/2$, and various pairs (m, n) in parentheses.

Figures 6.8a and 6.8b illustrate the *affine transformation* of the node grid circumscribed by $y/x = a/b = 1/1$ and $y/x = c/d = 1/2$ into the standard lattice with all of its prime nodes. Node $(3, 2)$ with respect to the x, y referential is the *affine transform* of node $(1, 1)$ with respect to the n, m referential, and vice versa. Similarly, node $(n, m) = (4, 1)$ is the affine transform of $(x, y) = (9, 5)$.

That notion may be extended to line slopes and their corresponding fractions, and we shall say that lines $m/n = 1/4$ and $y/x = 5/9$ are affine transforms, as are lines $y/x = 2/3$ and $n/m = 1/1$. The n-axis in Figure 6.8b corresponds to $m/n = 0/1$. It is the affine transform of line $y/x = c/d = 1/2$. Similary, the m-axis, which corresponds to $m/n = 1/0$, is the affine transform of $y/x = a/b = 1/1$.

The affine transforms of the x and y axes are not shown in Figure 6.8b. The reader may wish to draw them as an exercise. They form an obtuse angle inside which lies the n, m referential.

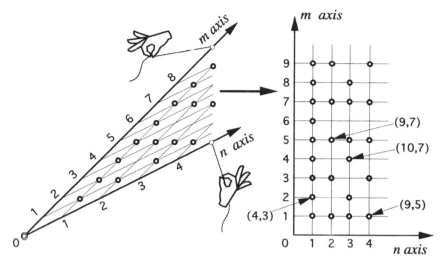

Figure 6.8b. The affine transformation. Pairs in parentheses are (x, y).

The Stern-Brocot Tree

The above discussion suggests the following exercise. Let us put

$$\frac{a}{b} = \frac{1}{0} \quad \text{and} \quad \frac{c}{d} = \frac{0}{1}, \tag{6.27}$$

and create the successive mediants of these two contiguous fractions:

Step 0. $\dfrac{1}{0}, \dfrac{0}{1}$

Step 1. $\dfrac{1}{0}, \dfrac{1}{1}, \dfrac{0}{1}$

Step 2. $\dfrac{1}{0}, \dfrac{2}{1}, \dfrac{1}{1}, \dfrac{1}{2}, \dfrac{0}{1}$ (6.28)

Step 3. $\dfrac{1}{0}, \dfrac{3}{1}, \dfrac{2}{1}, \dfrac{3}{2}, \dfrac{1}{1}, \dfrac{2}{3}, \dfrac{1}{2}, \dfrac{1}{3}, \dfrac{0}{1}$

. . .   . . .

Every step preserves the symmetry around fraction 1/1, on either side of which fall fractions y/x and x/y. We may thus confine our analysis to the construction's right side, which contains only fractional numbers, with the notable exception of 1/1.

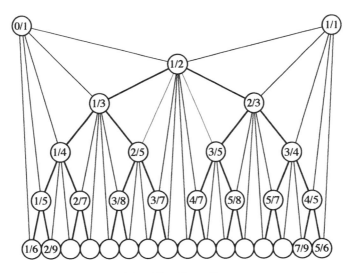

Figure 6.9a. The Stern-Brocot tree.

Step n brings with it 2^{n-1} new members. This again suggests that the successive generations are the branches of a binary tree, as in Figure 6.9a. That construction is known as the Stern-Brocot Tree. It was discovered independently by the German mathematician Moriz Stern[6] and the French clockmaker Achille Brocot.[7] The reader will easily fill in the empty bubbles. Each bubble contains an irreducible fraction smaller than its two predecessors, to which it is connected by a thin line and a thick line.

Figure 6.9b shows the projection upon a datum line, situated at infinity, of the rational numbers generated by the Stern-Brocot procedure. That datum line is of unit length, and each projection is made to fall at a distance from origin 0 equal to the fraction's value. The totality of nonnegative rational numbers less than 1 may thus be generated, with their projections succeeding one another in order of increasing value, confirming that no fraction can appear more than once upon the datum line.

Given the fractions $a/b > c/d$, what is the smallest denominator x for which there exists at least one relatively prime numerator

[6]Moriz Stern, "Über eine zahlentheoretische Funktion," *Journal für die reine und angewandte Mathematik* 55 (1858): 193–220.

[7]Achille Brocot, "Calcul des rouages par approximation, nouvelle méthode," *Revue Chronométrique* 6 (1860): 186–194.

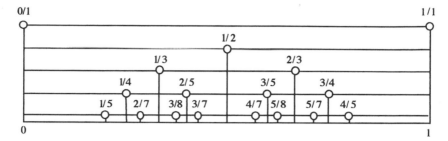

Figure 6.9b. Projection upon a datum line of the successive generations.

y such that $a/b > y/x > c/d$? Generally, solving indeterminate Diophantine equations almost invariably requires a search procedure, as in the following examples.

Consider, for example, the fractions $a/b = 5/13$ and $c/d = 1/6$. The sought-after denominator is 3, for which $5/13 > 1/3 > 1/6$. The reader may also verify that neither denominator 1 nor denominator 2 can solve the problem.

Generally, we may address that problem by turning to matrix equation (6.21a), namely,

$$\begin{bmatrix} m \\ n \end{bmatrix} = \begin{bmatrix} d & -c \\ -b & a \end{bmatrix} \times \begin{bmatrix} y \\ x \end{bmatrix},$$

plugging in increasing values of $x = 1, 2, 3, \ldots$, and then searching for the relatively prime values of $y = 1, 2, 3, \ldots$ that yield positive integral values for m, n. For this example, the matrix equation is

$$\begin{bmatrix} m \\ n \end{bmatrix} = \begin{bmatrix} 6 & -1 \\ -13 & 5 \end{bmatrix} \times \begin{bmatrix} y \\ x \end{bmatrix}.$$

Testing $x = 2$, we must solve $6y - 2 = m$ and $-13y + 10 = n$. The latter requirement cannot be satisfied.

Testing $x = 3$, we must solve $6y - 3 = m$ and $-13y + 15 = n$. The value $y = 1$ provides the solution ($m = 3, n = 2$), corresponding to $ma + nc = 17$ and $mb + nd = 51$, which, when divided by determinant $D = 17$, respectively yield $y = 1$, $x = 3$. (To understand why $ma + nc$ and $mb + nd$ must be divided by D in order to yield y and x, the reader should turn to matrix equation (6.21b).) In this example, there is no other value of y satisfying $5/13 > y/3 > 1/6$.

What can be said of the fraction 1/3 that we have just identified? It is certainly not "the smallest irreducible fraction larger than 1/6 and smaller than 5/13." There is no such thing! The fractions smaller than 1/3 and larger than 1/6 are infinite in number. What, then, is so special about 1/3? Well, it is the irreducible fraction smaller than 5/13 and larger than 1/6 with the smallest numerator and the smallest denominator. If we must absolutely coin a name for the prime node that possesses these characteristics, we might perhaps think of it as the *innermost prime node* larger than 1/6 and smaller than 5/3. Indeed, it is the prime node closest to the origin, "as the crow flies."

For one example, let us refer back to Figure 6.3. Inspection of the figure reveals that the innermost prime node larger than 3/5 and smaller than 1 is 2/3 in this case. A different example consists of choosing $a/b = 5/2$ and $c/d = 1/2$. To denominator $x = 1$ correspond the two irreducible fractions $y/x = 1/1$ and $y/x = 2/1$, both of which satisfy $5/2 > y/x > 1/2$. In this case, the innermost prime node is 1/1.

What these examples illustrate is that when the determinant of two given fractions is greater than 1, integers $m, n > 1$ may be found for which the innermost denominator is smaller than either b or d, and the innermost numerator is smaller than either a or c.

On the other hand, when $a/b \Rightarrow c/d$, the situation is crystal clear. The innermost prime node is none other than the mediant $(a + c)/(b + d)$, whose abscissa is obviously larger than the larger of b and d and whose ordinate is larger than the larger of a and c, in sharp contrast to what happens when considering noncontiguous fractions. Node $(x = b + d, y = a + c)$ is the affine transform of node $(n = 1, m = 1)$.

PENCILS AND LADDERS

Figure 6.10a represents an enlargement of the lower left corner of the $\sqrt{2}$ cleavage. Starting at $x = 1$, the bundle, or *pencil*, of lines containing $\sqrt{2}$ is bounded by (and contains) a lower limiting line, whose slope $\theta_{\sqrt{2}}(1) = h_{\sqrt{2}}(1)/1 = 1/1$ is referred to as the *floor* at abscissa x, and an upper limiting line (not contained in the pencil), whose slope $\Theta_{\sqrt{2}}(1) = (h_{\sqrt{2}}(1) + 1)/1 = 2/1$ is referred to as the *ceiling* at abscissa $x = 1$.

As we move on to $x = 2$, $h_{\sqrt{2}}(2) = 2$, corresponding to $\theta_{\sqrt{2}}(2) = 2/2 = 1/1$, and $\Theta_{\sqrt{2}}(2) = 3/2$. The floor thus remains

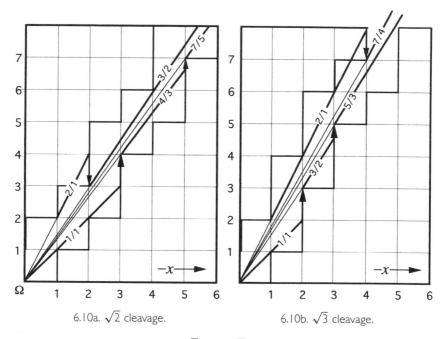

6.10a. $\sqrt{2}$ cleavage.   6.10b. $\sqrt{3}$ cleavage.

Figure 6.10. $\sqrt{2}$ and $\sqrt{3}$ cleavage pencils.

unchanged at 1/1, but the ceiling drops from 2/1 to 3/2. At $x = 3$, the floor rises to 4/3, while the ceiling remains at 3/2. As the value of x is increased, the ceiling can only drop or remain unchanged, and the floor can only rise or remain unchanged. Any such change will be referred to as an *event*. Clearly, events may only coincide with prime nodes. In Figure 6.10a, as we move beyond $x = 3$, the floor 4/3 flies above prime node 5/4, whose slope is smaller than 4/3, until it hits a vertical wall at $x = 5$, thus initiating the new 7/5 floor. Meanwhile, the ceiling 3/2 flies under prime node 5/3, and brushes nonprime node 6/4 without changing its course.

The floor and ceiling values at any given abscissa are said to be *concomitant*. For any value of x where no event occurs, the floor and ceiling are those of the two nearest previous events.

For example, for $\sqrt{2}$, the concomitant floor and ceiling at $x = 4$ are, respectively, 4/3 and 3/2. As x further increases, the pencil width progressively narrows with successive events, as its ceiling drops or its floor rises. As x goes to infinity, the pencil narrows down to a single line, whose slope is $\sqrt{2}$.

Figure 6.10b represents an enlargement of the lower left corner of the $\sqrt{3}$ cleavage.

Table 6.4 shows the *ladders*, that is, the successive different concomitant pairs, for $\sqrt{2}$ and $\sqrt{3}$.

TABLE 6.4
Ladders, showing floors and ceilings, for $\sqrt{2}$ and $\sqrt{3}$.

n	$\sqrt{2}$ c_n/d_n	a_n/b_n	$\sqrt{3}$ c_n/d_n	a_n/b_n
0	0/1	1/0	0/1	1/0
1	1/1	1/0 ⌉	1/1	1/0 ⌉
2	⌈ 1/1	2/1	⌈ 1/1	2/1
3	⌊ 1/1	3/2	3/2	2/1 ⌉
4	4/3	3/2 ⌉	5/3	2/1 ⌋
5	7/5	3/2 ⌋	⌈ 5/3	7/4
6	⌈ 7/5	10/7	12/7	7/4 ⌉
7	⌊ 7/5	17/12	19/11	7/4 ⌋
8	24/17	17/12 ⌉	⌈ 19/11	26/15
9	41/29	17/12 ⌋	45/26	26/15 ⌉
10	⌈ 41/29	58/41	71/41	26/15 ⌋
11	⌊ 41/29	99/70	⌈ 71/41	97/56
12	140/99	99/70 ⌉	168/97	97/56 ⌉
13	239/169	99/70 ⌋	265/153	97/56 ⌋
14	⌈ 239/169	338/239	⌈ 265/153	362/209
15	⌊ 239/169	577/408	627/362	362/209 ⌉
16	816/577	577/408 ⌉	989/571	362/209 ⌋
17	1393/985	577/408 ⌋	⌈ 989/571	1351/780
18	⌈ 1393/985	1970/1393	2340/1351	1351/780 ⌉
19	⌊ 1393/985	3363/2378	3691/2131	1351/780 ⌋
20	4756/3363	3363/2378 ⌉	⌈ 3691/2131	5042/2911
21	8119/5741	3363/2378 ⌋	8733/5402	5402/2911 ⌉
22	⌈ 8119/5741	11482/8119		⌋
...	⌊		...	

How are the ladders constructed? Suppose we wish to construct the ladder for the finite nonnegative number μ. We proceed as follows:

1. For $n = 0$, the floor $0/1$ is placed at the head of the left-hand column, and the ceiling $1/0$ at the head of the right-hand column.
2. Given the fractions a_n/b_n and c_n/d_n upon any rung $n = 1, 2, \ldots$, the elements of rung $n + 1$ are determined as follows:

- Put $y_{n+1} = a_n + c_n$, and $x_{n+1} = b_n + d_n$. $\hfill (6.29a)$

- If $\dfrac{y_{n+1}}{x_{n+1}} > \mu$, put $\dfrac{a_{n+1}}{b_{n+1}} = \dfrac{y_{n+1}}{x_{n+1}}$ and $\dfrac{c_{n+1}}{d_{n+1}} = \dfrac{c_n}{d_n}$. $\hfill (6.29b)$

- If $\dfrac{y_{n+1}}{x_{n+1}} \le \mu$, put $\dfrac{c_{n+1}}{d_{n+1}} = \dfrac{y_{n+1}}{x_{n+1}}$ and $\dfrac{a_{n+1}}{b_{n+1}} = \dfrac{a_n}{b_n}$. $\hfill (6.29c)$

The above construction rules are validated as follows. Upon any given rung n, the pencil surrounding μ consists of an infinite bundle of lines whose slope is equal to or larger than the floor c_n/d_n and smaller than the ceiling a_n/b_n. The next event, whether it is a drop of the ceiling or a rise of the floor, will occur upon the innermost prime node within that pencil.

For $n = 0$, the floor $0/1$ and ceiling $1/0$ verify

$$a_0/b_0 \Rightarrow c_0/d_0. \hfill (6.30)$$

Let us now jump to rung n and assume that

$$\frac{a_n}{b_n} \Rightarrow \frac{c_n}{d_n}. \hfill (6.31a)$$

The sought-after innermost prime node is $(x = b_n + d_n, y = a_n + c_n)$, corresponding to the median of the two fractions. Abiding by construction rules (6.29a)–(6.29c), we get either

$$\frac{a_{n+1}}{b_{n+1}} = \frac{a_n + c_n}{b_n + d_n} \quad \text{and} \quad \frac{c_{n+1}}{d_{n+1}} = \frac{c_n}{d_n},$$

or

$$\frac{c_{n+1}}{d_{n+1}} = \frac{a_n + c_n}{b_n + d_n} \quad \text{and} \quad \frac{a_{n+1}}{b_{n+1}} = \frac{a_n}{b_n},$$

and in either case, we have

$$\frac{a_{n+1}}{b_{n+1}} \Rightarrow \frac{c_{n+1}}{d_{n+1}}. \hfill (6.31b)$$

It follows by induction that the process is valid for any value of n. Concomitant fractions are always contiguous.

Beginning with the rung corresponding to $b = 2$, note that every floor subsumes all previous floors. The significance of the brackets in Table 6.4 and other tables will become apparent upon discussing the relationship between ladders and continued fractions. The numbers $\sqrt{2}$ and $\sqrt{3}$ are irrational. Their ladders go on forever, with the median alternatively shifting from right to left, and vice-versa.

Cleavages and Continued Fractions

In Tables 6.4, brackets were placed in the floor and ceiling columns according to the following rule:

1. Start at origin $n = 0$, and proceed downwards.
2. If two or more identical fractions appear in succession in any given column, place a bracket around them, with the exception of the first.

If the bracketed fractions in Table 6.4 are listed only once in their respective columns, Table 6.5 is obtained.

TABLE 6.5
Convergents for $\sqrt{2}$ and $\sqrt{3}$.

$\sqrt{2}$		$\sqrt{3}$	
Floor	*Ceiling*	*Floor*	*Ceiling*
	(1/0)		(1/0)
1/1		1/1	
	3/2		2/1
7/5		5/3	
	17/12		7/4
41/29		19/11	
	99/70		26/15
239/169		71/41	
	577/408		97/56
1393/985		265/153	
	3363/2378		362/209
. . .			

We observe that for $\sqrt{2}$

$$\frac{1}{0} \Rightarrow \frac{1}{1} \Leftarrow \frac{3}{2} \Rightarrow \frac{7}{5} \Leftarrow \frac{17}{12} \Rightarrow \frac{41}{29} \Leftarrow \frac{99}{70} \Rightarrow \frac{239}{169} \Leftarrow \frac{577}{408} \cdots$$

and for $\sqrt{3}$

$$\frac{1}{0} \Rightarrow \frac{1}{1} \Leftarrow \frac{2}{1} \Rightarrow \frac{5}{3} \Leftarrow \frac{7}{4} \Rightarrow \frac{19}{11} \Leftarrow \frac{26}{15} \Rightarrow \frac{71}{41} \Leftarrow \frac{97}{56} \cdots$$

With the exception of 1/0, these fractions exactly correspond to the succesive convergents $\partial_0, \partial_1, \partial_2 \ldots$ of the continued fraction

$$\mu = q_0 + \cfrac{1}{q_1 + \cfrac{1}{q_2 + \cfrac{1}{q_3 + \cfrac{1}{\cdots}}}}$$

where the quotients $q_0, q_1, q_2, q_3 \ldots$ are none other than the number of rows encompassed by the successive brackets, beginning with the group of rows at the top of the right-hand column. Thus,

$$\sqrt{2} = 1 + \cfrac{1}{2 + \cfrac{1}{2 + \cfrac{1}{2 + \cfrac{1}{\cdots}}}}$$

and

$$\sqrt{3} = 1 + \cfrac{1}{1 + \cfrac{1}{2 + \cfrac{1}{1 + \cfrac{1}{2 + \cfrac{1}{\cdots}}}}}$$

The reason for this kinship between ladders and continued fractions is best understood with the help of the additional example of Table 6.6a, which is constructed for π, and the accompanying diagram

TABLE 6.6a
Ladder for π.

	c_n/d_n	a_n/b_n	
	0/1	1/0	
	1/1	1/0 ⎤	
	2/1	1/0	$q_0 = 3$ rungs
	3/1	1/0 ⎦	
	⎡ 3/1	4/1	
	⎢ 3/1	7/2	
	⎢ 3/1	10/3	
$q_1 = 7$ rungs	⎢ 3/1	13/4	
	⎢ 3/1	16/5	
	⎢ 3/1	19/6	
	⎣ 3/1	22/7	
	25/8	22/7 ⎤	
	47/15	22/7	
	69/22	22/7	
	91/29	22/7	$q_2 = 15$ rungs
	...		
	311/99	22/7	
	333/106	22/7 ⎦	
$q_3 = 1$ rung	⎡ 333/106	355/113	
	688/219	355/113 ⎤	
	1043/332	355/113	
	1398/445	355/113	
	...		$q_4 = 292$ rungs
	103283/32876	355/113	
	103638/32989	355/113	
	103993/33102	355/113 ⎦	
$q_5 = 1$ rung	⎡ 103993/33102	104348/33215	
	208341/66317	104348/33215 ⎦	$q_6 = 1$ rung
	...		

of Table 6.6b. From Table 6.6a, we obtain the continued fraction

$$\pi = 3 + \cfrac{1}{7 + \cfrac{1}{15 + \cfrac{1}{1 + \cfrac{1}{292 + \cfrac{1}{1 + \cdots}}}}}.$$

TABLE 6.6b
From pencils to convergents, where
N = numerator, and D = denominator.

c_n/d_n	a_n/b_n	
	N_0/D_0	
	N_0/D_0	
	$\cdots$	q_0 rungs
	N_0/D_0	
	$\cdots$	
$\cdots$		
N_{n-1}/D_{n-1}	$\cdots$	
N_{n-1}/D_{n-1}	N_n/D_n	
	N_n/D_n	
	$\cdots$	q_n rungs
N_{n+1}/D_{n+1}	N_n/D_n	
N_{n+1}/D_{n+1}	$\cdots$	
$\cdots$		

Referring to Table 6.6b, we have

$$N_n = N_{n-2} + q_n N_{n-1} \quad \text{and} \quad D_n = D_{n-2} + q_n D_{n-1},$$

with (6.32)

$$N_{-2} = 0, \qquad D_{-2} = 1, \qquad N_{-1} = 1, \qquad D_{-1} = 0.$$

These are the very recursion formulae governing the calculation of the convergent $\delta_n = N_n/D_n$ of a continued fraction.

The above continued fraction for π constitutes the first part of a much longer nonperidic continued fraction discovered by Johann Heinrich Lambert (1728–1777), who established the irrationality of π. Its transcendental character was established by Lindemann in 1882. The fraction's four initial convergents, according to Petr Beckman's account,[8] were discovered in one form or another through the ages:

> "3" was the value contained in the Old Testament verse "Also he made a molten sea of ten cubits from brim to brim, round the compass, and five cubits the height thereof; and a line of thirty cubits did compass it round about" (1 Kings 7:23);
>
> "22/7" is the upper bound of π calculated by Archimedes, using the method of exhaustion;
>
> "333/106" was the lower bound calculated by Adriaan Anthoniszoon, around 1583; and
>
> "355/113" was also found by Anthoniszoon, as well as by Viete and other sixteenth-century mathematicians.

Fractions 3/1 and 333/106 are floors, while 22/7 and 355/113 are ceilings. It must be noted that Archimedes provided the statement $223/71 < \pi < 22/7$, whose bounds are none other than the concomitant fractions on the seventeenth rung of the ladder in Table 6.6a! Surely, more precise exact statements may be made as one descends the ladder.

What happens when the number μ is rational? To answer that question, consider the the ladder in Table 6.6c, which was constructed for $\mu = 9/25$. As we go down the ladder's rungs, we follow instructions (6.29a)–(6.29c). The procedure may not go on indefinitely without encountering 9/25 at some stage, as we zero in on the required fraction from both ends. The fraction 9/25 remains unchanged forever on the left, while the fraction on the right monotonously decreases, tending to 9/25. This leads to the continued fraction

$$9/25 = [0, 2, 1, 3, 2, \infty]$$
$$= 1/[2, 1, 3, 2]$$

[8]Petr Beckman, *A History of π* (New York: St. Martin's Press, 1971).

TABLE 6.6c
Ladder for
generating 9/25.

c_n/d_n	a_n/b_n
0/1	1/0
⎡ 0/1	1/1
⎣ 0/1	1/2
1/3	1/2 ⎤
⎡ 1/3	2/5
1/3	3/8
⎣ 1/3	4/11
5/14	4/11 ⎤
9/25	4/11 ⎦
⎡ 9/25	13/36
9/25	22/61
⎣ 9/25	31/36
⋮	. . .

$$\downarrow$$
$$\infty$$

Klein's Construction

An alternative geometrical model was offered by Felix Klein[9] and described by H. Davenport.[10] That model is reinterpreted here in terms of cleavages (Figure 6.11). Two strings originate at infinity within the cleavage, and one of them is wrapped around whichever corner of the lower cleavage bank it encounters, as it is pulled downwards, while the other is wrapped upwards in similar fashion around the corners of the upper cleavage bank, as in Figure 6.11b, which was drawn for the Golden Section. Each string either bends around a corner or brushes against it without changing its course, unless it clears it altogether. The

[9]Felix Klein, *Ausgewählte Kapitel der Zahlentheorie* (Teubner, 1907), pp. 17–25.
[10]Harold Davenport, *The Higher Arithmetic* (New York: Harper, 1960).

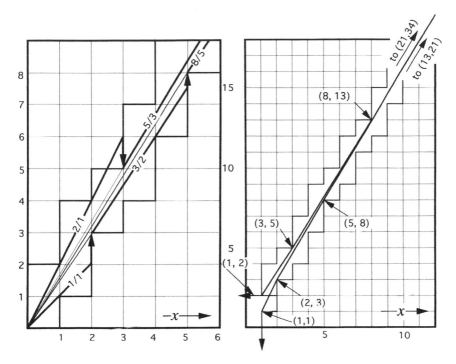

Figure 6.11a. Cleavage pencils.   Figure 6.11b. Klein's construction

Figure 6.11. Constructions for the Golden Section.

corners that bend the string are precisely those prime nodes that cor-respond to convergents of the continued fraction. Figure 6.11a shows the Golden Section's pencils.

The Greatest Common Divisor Revisited

Here, we offer a method for calculating the GCD of two numbers x, y, where $y > x$. That method exhibits a close kinship with Euclid's algorithm, but is different from it, and often requires a smaller number of steps. Let

$$a = [y/x] \qquad (6.33)$$

represent the integral part of y/x. In other words,

$$a + 1 > \frac{y}{x} > a, \tag{6.34}$$

where a is an integer. Also let

$$k = y - ax \quad \text{and} \quad l = (a + 1)x - y. \tag{6.35}$$

Any integer that divides x and y also divides both $k = (y - ax)$ and $l = (a + 1)x - y$. Additionally, if some integer divides A and $(A - B)$, it also divides B. Consequently, if some integer divides both $k = (y - ax)$ and $l = (a + 1)x - y = x - (y - ax)$, it also divides k and divides l. Hence

$$GCD(y, x) = GCD(k, l). \tag{6.36}$$

Additionally, it follows from (6.35) that

$$k < y. \tag{6.37a}$$

It also follows that

$$y > ax \therefore 0 > ax - y \therefore x > x + ax - y \therefore x > (a + 1)x - y;$$

in other words,

$$l < x. \tag{6.37b}$$

Similarly, it follows that

$$y/x < a + 1 \therefore y < ax + x \therefore y - ax < x;$$

in other words,

$$k < x. \tag{6.37c}$$

Finally, since $a \geq 1$, it follows that $2a \geq 1 + a$. And from (6.34),

$$y/x > a \therefore 2y/x > 2a \therefore 2y/x > 1 + a$$
$$\therefore 2y = (a+1)x \therefore y = (a+1)x - y$$

in other words,

$$l < y. \tag{6.37d}$$

In summary, $k < x$, $k < y$, $l < x$, $l < y$. Starting with fraction y/x, the exercise thus boils down to generating the fraction k/l as in 6.34 and 6.35, then substituting k/l for y/x and generating k', l'; and so on, until the process halts. In order for the fractions examined at every step to be larger than 1, it is sometimes necessary to swap k and l.

Example 6.5

Find GCD(107 415, 85 911). We proceed as follows:

1.
$$2 > \frac{107{,}415}{85{,}911} > 1$$
$$k = 107{,}415 - 85{,}911 = 21{,}504,$$
$$l = 2 \times 85{,}911 - 107{,}415 = 64{,}407$$

2.
$$3 > \frac{64{,}407}{21{,}504} > 2$$
$$k' = 64{,}407 - 2 \times 21{,}504 = 21{,}399,$$
$$l' = 3 \times 21{,}504 - 64{,}407 = 105$$

3.
$$204 > \frac{21{,}399}{105} > 203$$
$$k'' = 21{,}399 - 203 \times 105 = 84,$$
$$l'' = 204 \times 105 - 21{,}399 = 21$$

Since 84 is divisible by 21, we get GCD(107 415, 85 911) $= 21$. The above example does not fundamentally differ much from Euclid's algorithm.

Example 6.6

Obtain the GCD(1,181,565, 85,911). We proceed as follows:

$$14 > 1,181,565/85,911 > 13$$
$$14 > \quad 64,722/21,189 \quad > 13$$
$$18 > \quad 20,034/1,155 \quad > 17$$
$$2 > \quad\quad 756/399 \quad\quad > 1$$
$$9 > \quad\quad 357/42 \quad\quad > 8$$
$$21/21 \quad\quad = 1$$

Hence GCD(1 181 565, 85 911) = 21. In this case, Euclid's algorithm would proced in eight steps, as follows:

$$1,181,565 = 85,911 \times 13 + 64,722$$
$$85,911 = 64,722 \times 1 + 21,189$$
$$64,722 = 21,189 \times 3 + 1,155$$
$$21,189 = 1,155 \times 8 + 399$$
$$1,155 = 399 \times 2 + 357$$
$$399 = 357 \times 1 + 42$$
$$357 = 42 \times 8 + 2$$
$$42 = 21 \times 2 + 0$$

For obvious reasons, many identical numbers may be encountered in both procedures. The first was nonetheless somewhat faster than the second.

MARGINALIA

Cleavages and Positional Number Systems

Positional number systems offer a striking illustration of the definition of real numbers in terms of an infinite set of coherent nodes, when the chosen abscissas are precisely the consecutive "weights"

$\pi_0, \pi_1, \pi_2, \ldots$ corresponding to some base, be it mixed or uniform. Given base b and the number μ, the infinite sequence

$$S(\mu)_b = \left\{ \left(\pi_0, h_\mu(\pi_0)\right), \left(\pi_1, h_\mu(\pi_1)\right), \left(\pi_2, h_\mu(\pi_2)\right), \ldots \right\} \quad (6.38)$$

will be referred to as the *base b cleavage sequence for* π.

Examples

For $\mu = \sqrt{2} - 1$, $b = 10$, the sequence is[11]

$$S(\sqrt{2} - 1)_{10} = \left\{ (1, 0), (10, 4), (100, 41), \right.$$
$$\left. (1000, 414), (10000, 4142), \ldots \right\}.$$

That set uniquely defines $\sqrt{2} - 1$. Writing $h_{\sqrt{2}-1}(x)$ in binary form, and using bold characters to distinguish binary from decimal notation, we get

$$S(\sqrt{2} - 1)_2 = \left\{ (1, \mathbf{0}.), (2, \mathbf{0}.), (4, \mathbf{1}.), (8, \mathbf{11}.), (16, \mathbf{110}.), \right.$$
$$\left. (32, \mathbf{1101}.), (64, \mathbf{11010}.), \ldots \right\}.$$

The factorial base cleavage sequence of e's mantissa is

$$S(e - 2)_f = \left\{ (1!, 0), (2!, 1), (3!, 4), (4!, 17), (5!, 86), \ldots \right\},$$

and if we define the companion bases of the factorial base as

$$f_0 = (\ldots 5, 4, 3, 2 . 2, 3, 4, 5, 6, \ldots),$$
$$f_{-1} = (\ldots 5, 4, 3, 2, 2 . 3, 4, 5, 6, \ldots),$$
$$f_{-2} = (\ldots 5, 4, 3, 2, 2, 3 . 4, 5, 6, \ldots),$$
$$f_{-3} = (\ldots 5, 4, 3, 2, 2, 3, 4 . 5, 6, \ldots),$$
$$f_{-4} = (\ldots 5, 4, 3, 2, 2, 3, 4, 5 . 6, \ldots),$$
$$\ldots,$$

---

[11]Unless otherwise specified, the numbers in parentheses are in decimal notation.

we may write

$$h_{(e-2)}(\pi_0) = h_{(e-2)}(1) \quad = (0.)_{f_0} \quad = 0,$$

$$h_{(e-2)}(\pi_1) = h_{(e-2)}(2) \quad = (1.)_{f_{-1}} \quad = 1,$$

$$h_{(e-2)}(\pi_2) = h_{(e-2)}(6) \quad = (11.)_{f_{-2}} \quad = 4,$$

$$h_{(e-2)}(\pi_3) = h_{(e-2)}(24) \quad = (111.)_{f_{-3}} \quad = 17,$$

$$h_{(e-2)}(\pi_4) = h_{(e-2)}(120) = (1111.)_{f_{-4}} = 86,$$

$$\ldots,$$

and we get the following expression for the factorial base cleavage sequence of $(e-2)$:

$$S(e-2)_f = \Big\{ (1, (0.)_{f_0}), (2, (1.)_{f_{-1}}), (6, (11.)_{f_{-2}}),$$
$$(24, (111.)_{f_{-3}}), \ldots \Big(\pi_n, (\overset{\overleftarrow{\quad n \quad}}{111\ldots1}.)_{f_{-n}} \Big), \ldots \Big\}.$$

Considering the base b cleavage subsets in the previous examples, it is clear that every node in the subset subsumes every preceding node. Only the last available node in any such enumeration is therefore significant, as it renders every preceding node redundant.

In the chapter on positional number systems, we learned that the mantissa of any (nonnegative) fractional number μ in a reflected mixed base (one for which $m_{-i} = m_{i-1}$) is such that

$$\delta^{\mu}_{-i} = h_{\mu}(\pi_i) - m_i h_{\mu}(\pi_{i-1}), \qquad (6.39)$$

where δ^{μ}_{-i} is the $-i$th digit of μ, and $\pi_i = m_0 m_1 m_2 \ldots m_{i-1}$, with $\pi_0 = 1$. The geometric interpretation of equation (6.39) is illustrated by Figures 6.12a and 6.12b, which were, respectively, drawn for $\sqrt{2} - 1$ in base 2 and $(e - 2)$ in the factorial base. If the straight line joining Ω and node $(\pi_n, h_{\mu}(\pi_n))$ is extended until it intersects the vertical line whose abscissa is π_{n+1}, and h denotes the intersection point's ordinate, the difference between the ordinates $h_{\mu}(\pi_{n+1})$ and h is none other than to ∂_{n+1}.

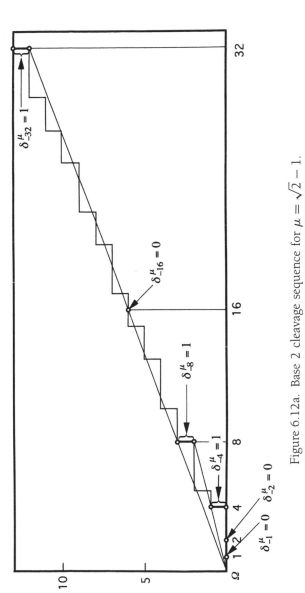

Figure 6.12a. Base 2 cleavage sequence for $\mu = \sqrt{2} - 1$.

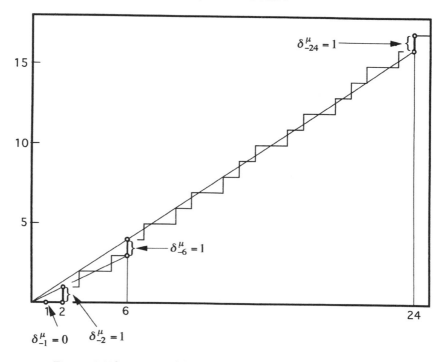

Figure 6.12b. Factorial base cleavage sequence for $\mu = e - 2$.

Cleavages and Automata

Imagine an automaton with one input port and one output port (Figures 6.13a and 6.13b). When an integer is fed into the automaton's input port, a corresponding finite sequence flows out of the output port, according to some specified rules. An appropriate device then instructs the output sequence to be fed back into the automaton. Every integer in that sequence is treated according to the same specified rules, and a new sequence flows out. The process is repeated indefinitely, resuting in an output sequence of infinite length.

Example 6.7

Automaton for the generation of H_ϕ, $\phi = \dfrac{1 + \sqrt{5}}{2}$

    Input integer $= 1 \rightarrow$ Output sequence $= 2$
    Input integer $= 2 \rightarrow$ Output sequence $= 21$
    Initial input $= 2$

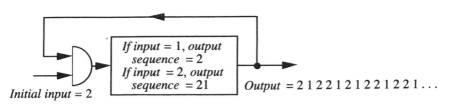

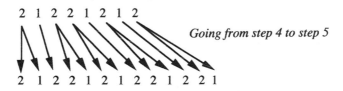

Going from step 4 to step 5

		Number of digits	Sum of digits
Step 0	2	1	2
Step 1	21	2	3
Step 2	212	3	5
Step 3	21221	5	8
Step 4	21221212	8	13
Step 5	2122121221221	13	21
. . .			

Figure 6.13a. Automaton for the generation of H_φ, where $\phi = (1 + \sqrt{5})/2$.

The ratio (sum)/(number) of digits converges to the Golden Section $(1 + \sqrt{5})/2$. Both sequences are identical to the Fibonacci sequence.

Example 6.8

Automaton for the generation of $H_{\sqrt{2}}$

> Input integer $= 1 \rightarrow$ Output sequence $= 12$
> Input integer $= 2 \rightarrow$ Output sequence $= 121$
> Initial input $= 1$

The ratio (sum of digits)/(number of digits) at step n is none other than convergent ∂_n of the continued fraction corresponding to $\sqrt{2}$. It will be observed that the initial portion of any sequence consists of the previous sequence, and if an additional 1 is placed at the beginning, the sequence $H_{\sqrt{2}}$ is obtained.

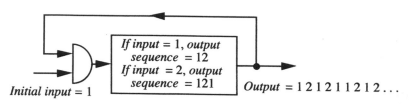

	Number of digits	Sum of digits
Step 0 1	1	1
Step 1 12	2	3
Step 2 12121	5	7
Step 3 121211212112	12	17
Step 4 12121121211212121121211212121	29	41
...		

Figure 6.13b. Automaton for the generation of $H_{\sqrt{2}}$.

Example 6.9

Automation for the generation of $H_{\sqrt{3}}$

Input integer $= 1 \rightarrow$ Output sequence $= 221$
Input integer $= 2 \rightarrow$ Output sequence $= 2212$
Initial input 0 : 1

The reader will verify that, upon placing an additional 1 at the beginning of the output sequence, the sequence of Table 6.2 is obtained, corresponding to $\sqrt{3}$.

Cleaving Crystals

In an article on quasi-crystals, Peter Stephens and Alan Goldman examined the problem of shearing a cubic lattice at a certain angle, when that angle's tangent is rational and when it is irrational.[12] The top left panel of Plate 10 shows a cube with three different cleaving angles, each of which generates one of the subsequent diagrams. The top right panel corresponds to a diagonal cut on each of three adjacent faces. The increment sequence is 1111111..., as one climbs

---

[12]Peter Stephens and Alan Goldman, "La Structure des quasi-cristaux," *Pour la science*, June 1991: 56–63.

the ladder in the northeast or northwest direction. The bottom left panel corresponds to a slope of 4/3 upon the right-hand face, and 2/3 upon the left-hand face. The corresponding increment sequences are 112112112 . . . climbing northeast, and 011011011 . . . climbing northwest. The bottom right panel corresponds to an irrational angle on both faces. The sequence lengths are too short for a reasonable estimation of the slopes, which seem to approach 6/5 in one direction, and 5/8 in the other.

If these diagrams are regarded not as isometric views of three-dimensional objects, but as plane patterns, the little white parallelograms in the top right and bottom left panels are capable of *regularly tiling* the plane following translations parallel to their respective edges. The bottom right panel contains no such elementary cell, and the floor tiling is irregular. Tiling was extensively studied by Roger Penrose, who discovered a celebrated method for irregularly tiling the plane using two complementary portions of an isoceles triangle whose base angle is $360°/5 = 72°$ or $180°/5 = 36°$. The ratio of that triangle's long to its short side is the quadratic irrational $(1 + \sqrt{5})/2$, known as the *Golden Number*, which has received an enormous amount of quasi-mystical attention over the centuries.

Cleavages and Replicative Functions

A function f(x) such that

$$f(nx) = f(x) + f\left(x + \frac{1}{n}\right) + f\left(x + \frac{2}{n}\right) + \cdots + f\left(x + \frac{n-1}{n}\right),$$

where n is any positive integer, is called a *replicative function*. Examples of replicative functions may be found in Donald E. Knuth's *Fundamental Algorithms*,[13] such as

1. $f(x) = x - 1/2$.
2. $f(x) = 1$ if x is an integer, and 0 otherwise.
3. $f(x) = \log |2 \sin \pi x|$ if the value $f(x) = -\infty$ is allowed.

Let us now consider the fractional number $0 \leq \nu < 6/5$, and compute the values of $[\mu]$, $[\mu + 1/5]$, $[\mu + 2/5]$, $[\mu + 3/5]$, $[\mu + 4/5]$ for those values of μ contained within the successive ranges

[13]Donald E. Knuth, *Fundamental Algorithms* (New York: Addison-Wesley, 1969).

TABLE 6.7
Replicative functions.

μ	$[\mu]$	$[\mu + 1/5]$	$[\mu + 2/5]$	$[\mu + 3/5]$	$[\mu + 4/5]$	$[5\mu]$
$0 \leq \mu < 1/5$	0	0	0	0	0	0
$1/5 \leq \mu < 2/5$	0	0	0	0	1	1
$2/5 \leq \mu < 3/5$	0	0	0	1	1	2
$3/5 \leq \mu < 4/5$	0	0	1	1	1	3
$4/5 \leq \mu < 1$	0	1	1	1	1	4
$1 \leq \mu < 6/5$	1	1	1	1	1	5

$0 \leq \mu < 1/5$, $1/5 \leq \mu < 2/5$, $2/5 \leq \mu < 3/5, \ldots, 1 \leq \mu < 6/5$. This computation, shown in Table 6.7, allows us to write

$$[5\mu] = [\mu] + [\mu + 1/5] + [\mu + 2/5] + [\mu + 3/5] + [\mu + 4/5].$$

If μ represents a positive real number, and x any positive integer, the above may be generalized as follows:

$$h_\mu(x) = [\mu x] = [\mu] + [\mu + 1/x] + [\mu + 2/x] + \cdots + [\mu + (x-1)/x].$$

For example,

$$h_{\sqrt{2}}(5) = [\sqrt{2}] + [\sqrt{2} + 1/5] + [\sqrt{2} + 2/5] + [\sqrt{2} + 3/5] + [\sqrt{2} + 4/5]$$
$$= 1 + 1 + 1 + 2 + 2 = 7.$$

The interested reader may discover particular virtues to cleavage height $h_\mu(x)$, when expressed as above.

Gaussian Primes

The reader who is familiar with Gaussian primes will have recognized the kinship between the prime node pattern and the representation upon the Argand diagram of the complex numbers defined by Gauss as prime.

Complex number $(3 + i)$ may be expressed as $(2 - i)(1 + i)$, whereas neither $(2 - i)$ nor $(1 + i)$ may be expressed as products of complex numbers other than themselves and integer $1 = (1 + i \times 0)$. Complex numbers $(2 - i)$ and $(1 + i)$ are referred to as Gaussian primes. According to that definition, $(3 + i)$ is composite, in other words, not prime. With the exception of numbers of the form $(d, 0)$ and $(0, d)$, that is, those lying on the horizontal and vertical axes, Gaussian primes are a subset of the prime nodes of the number lattice. Gauss showed that his primes were of one of the following three forms (Figure 6.14):

1. $\pm 1 \pm i$ (black circles in the figure)
2. $\pm p$ and $\pm ip$, where p is a prime number of the form $4n + 3$ (square nodes)
3. $(\pm a \pm ib)$ and $(\pm b \pm ia)$, where a and b are nonzero and different and $a^2 + b^2$ is prime (white circles)

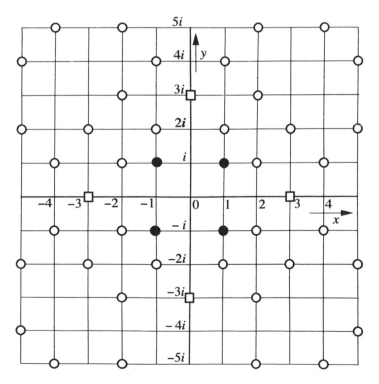

Figure 6.14. Gaussian primes.

The Gaussian primes lying on the horizontal axis are obviously prime in the traditional sense; in other words, they are prime integers. These may not be expressed as products of integer primes or Gaussian primes. On the other hand, certain prime integers, such as 5 and 29, may be expressed as products of Gaussian primes. Thus $5 = (2 + i)(2 - i)$ and $29 = (5 + 2i)(5 - 2i)$. Gauss showed that prime integers of the form $4n + 1$ may uniquely be factored into Gaussian primes.

APPENDIX 6.1

Proof of Test (6.7)

Number e is defined as

$$e = 1 + \frac{1}{1!} + \frac{1}{2!} + \frac{1}{3!} + \cdots.$$

It is desired to prove that for any positive integer x,

$$[xe] = \left[x \left(1 + \frac{1}{1!} + \frac{1}{2!} + \frac{1}{3!} + \cdots + \frac{1}{x!} \right) \right].$$

Putting

$$R = \left(\frac{1}{(x+1)!} + \frac{1}{(x+2)!} + \cdots \right),$$

that is,

$$e = 1 + \frac{1}{1!} + \frac{1}{2!} + \frac{1}{3!} + \cdots + \frac{1}{x!} + R,$$

and mutiplying throughout by x, it follows that

$$x \left(1 + \frac{1}{1!} + \frac{1}{2!} + \cdots + \frac{1}{x!} \right) < xe = x \left(1 + \frac{1}{1!} + \frac{1}{2!} + \cdots + \frac{1}{x!} \right) + xR.$$

Turning to R, we may write

$$x!R = \left(\frac{x!}{(x+1)!} + \frac{x!}{(x+2)!} + \cdots \right)$$

$$= \left(\frac{1}{(x+1)} + \frac{1}{(x+1)(x+2)} + \frac{1}{(x+1)(x+2)(x+3)} + \cdots \right).$$

Comparing the above expression term for term with

$$\left(\frac{1}{(x+1)} + \frac{1}{(x+1)^2} + \frac{1}{(x+1)^3} + \cdots \right) = \frac{1}{x},$$

we obtain $x!R < \dfrac{1}{x}$, that is,

$$xR < \frac{1}{x!}.$$

Hence

$$x\left(1 + \frac{1}{1!} + \frac{1}{2!} + \cdots + \frac{1}{x!} \right) < xe < x\left(1 + \frac{1}{1!} + \frac{1}{2!} + \cdots + \frac{1}{x!} \right) + \frac{1}{x!}.$$

We now show that for any value of integer x in the following inequality, p may not be an integer.

$$x\left(1 + \frac{1}{1!} + \frac{1}{2!} + \cdots + \frac{1}{x!} \right) < p < x\left(1 + \frac{1}{1!} + \frac{1}{2!} + \cdots + \frac{1}{x!} \right) + \frac{1}{x!}.$$

Multiplying the above statement throughout by $x!$ and putting

$$x\left(x! + \frac{x!}{1!} + \frac{x!}{2!} + \frac{x!}{3!} + \cdots + \frac{x!}{x!} \right) = q,$$

where q is an integer, we get $q < px! < q + 1$, which signifies that p may not be an integer. Now that we have established that the interval to which xe belongs may not contain an integer, the largest integer beneath that interval is

$$[xe] = \left[x \left(1 + \frac{1}{1!} + \frac{1}{2!} + \frac{1}{3!} + \cdots + \frac{1}{x!} \right) \right].$$

Q.E.D.

APPENDIX 6.2

The Increment Sequence[14]

Let ν be the fractional number ($0 \leq \nu < 1$) whose cleavage line is being constructed. Imagine that as you move along the x-axis, you come upon some abscissa b for which $\partial_\nu(b) = 1$, meaning that the value of νb has either just exactly reached (which may happen only if ν is rational) or just crossed some new integral value $[\nu b]$ on its way up. In the latter case, it may overshoot that value only by a quantity less than ν. In other words, $\nu b = [\nu b] + \partial$, with $0 \leq \partial < \nu$. The "cushion" c left between νb and the next integer $[\nu b] + 1$ is such that $1 \geq (c = 1 - \partial) > 1 - \nu$. We now examine three cases:

i. $0 < \nu < .5$

Let n be the largest integer such that $\nu n < 1$. If νn is smaller than the cushion, the next value of x for which $\partial_\nu(x) = 1$ is $b + n + 1$. If νn is equal to or larger than the cushion, that value is $b + n$. The increment sequence thus consists of packets of n zeros and packets of $(n - 1)$ zeros separated by single ones, or simply packets of n

---

[14]I am grateful to Professor Donald E. Knuth for the following comments: "Your discussion of cleavages relates to a subject that has been studied and discussed by many mathematicians; for example, Ivan Niven calls the numbers $h_\mu(x)$ the spectrum of μ in his nice book *Diophantine Approximations* (1963). The sequences you call Δ_μ have remarkable properties discovered by A. A. Markov and others; see, for example, articles by Stolarsky in *Canadian Math. Bulletin* **19** (1976), 473ff., and Fraenkel in *Trans. Amer. Math. Soc.* **341** (1994), 639ff. What you call "elementary ladders" can be traced to Daniel Schwenter in 1636; Wallis discussed them [in] Chapter[s] 10 and 11 of his *Algebra* (1985), and Christiaan Huyghens used the ideas for the gear wheels of his planetarium (c 1700)" (personal communication).

zeros separated by single ones when $1/\nu$ is an integer. For example,

$$\nu = 3/10 : n = 3$$

$$\Delta_{3/10} = 0\ 0\ 0\ 1\ 0\ 0\ 1\ 0\ 0\ 1, 0\ 0\ 0\ 1\ 0\ 0\ 1\ 0\ 0\ 1, \ldots$$

The increment sequence is periodic. Each cycle is of length 10 and contains three ones.

$$\nu = 1/3 : n = 2 \qquad \Delta_{1/3} = 0\ 0\ 1, 0\ 0\ 1, 0\ 0\ 1 \ldots$$

The increment sequence is periodic. Each cycle is of length 3 and contains one one.

$$\nu = \sqrt{2} - 1 : n = 2$$

$$\Delta_{\sqrt{2}-1} = 0\ 0\ 1\ 0\ 1\ 0\ 0\ 1\ 0\ 1\ 0\ 0\ 1\ 0\ 1\ 0\ 1\ 0\ 0\ 1 \ldots$$

The increment sequence is not periodic. The proportion of ones within any given chunk of consecutive increments approaches $\sqrt{2} - 1$ as the chunk gets larger. In the above sequence, that proportion is 4/10, which constitutes a rough approximation of $\sqrt{2} - 1$.

ii. $\nu = 1 - \nu = .5$

The increment sequence is $\Delta_{0.5} = 0\ 1\ 0\ 1\ 0\ 1\ 0\ 1\ 0\ 1\ 0\ 1 \ldots$

iii. $.5 < \nu < 1$.

If n is the largest integer such that $n(1 - \nu) < 1$, it may be shown that the increment sequence consists of packets of n ones and packets of $(n - 1)$ ones separated by single zeros, or packets of n ones separated by single zeros when $1/(1 - \nu)$ is an integer. For example,

$$\nu = 7/10 : n = 3 \qquad \Delta_{7/10} = 0\ 1\ 1\ 0\ 1\ 1\ 0\ 1\ 1\ 1, 0\ 1 \ldots$$

The increment sequence is periodic. As in the case of $\nu = 3/10$, each cycle is of length 10, but contains seven ones instead of three ones.

$$\nu = 2/3 : n = 2 \qquad \Delta_{2/3} = 0\ 1\ 1, 0\ 1\ 1, 0\ 1\ 1 \ldots$$

The increment sequence is periodic. As in the case of $\nu = 1/3$, each cycle is of length 3, but contains two ones instead of one one.

$$\nu = \sqrt{3} - 1 : n = 3$$

$$\Delta_{\sqrt{3}-1} = 0\ 1\ 1\ 0\ 1\ 1\ 1\ 0\ 1\ 1\ 1\ 0\ 1\ 1\ 0\ 1\ 1\ 1\ 0\ 1\ 1\ 1\dots$$

The proportion of ones in the above nonperiodic sequence chunk is $16/22 \approx .7272$, to be compared to $\sqrt{3} - 1 \approx .732\dots$.

 **CHAPTER 7**

Infinity

> When we've been there ten thousand years,
> Bright shining as the sun,
> We've no less days to sing God's praise
> Than when we first begun.
> ("Amazing Grace")

The road to infinity is strewn with paradoxes, and extreme care must be exercised when extrapolating one's reasoning from *here* to *there*. What may seem a natural extension of what we do within the realm of our immediate reach, in other words, with the first few integers, may sometimes lead to insurmountable contradictions when dealing with infinity, confirming the admission by one of the characters in Racine's play *Les Plaideurs*, when he declared *"Ce que je sais le mieux, c'est mon commencement"* (What I know best is my beginning).

CONVERGENCE

> "Can you do addition?" the White Queen said. "What's one and one and one and one and one and one and one and one and one and one?" "I don't know," said Alice. "I lost count."
> (Lewis Carroll, *Alice in Wonderland*)

Consider the arbitrary base

$$b = \left(m_{L-1}, m_{L-2}, \ldots, m_1, m_0 . m_{-1}, m_{-2} \ldots m_{-R}\right). \quad (7.1)$$

The largest expressible mantissa of length R is

$$\sum_{i=-1}^{-R} (m_i - 1)\pi_i$$

$$= \frac{m_{-1} - 1}{m_{-1}} + \frac{m_{-2} - 1}{m_{-1} - m_{-2}} + \cdots + \frac{m_{-R} - 1}{m_{-1}m_{-2}\cdots m_{-R}}$$

$$= 1 + \left(-\frac{1}{m_{-1}} + \frac{1}{m_{-1}}\right) + \left(-\frac{1}{m_{-1}m_{-2}} + \frac{1}{m_{-1}m_{-2}}\right) + \cdots$$

$$- \frac{1}{m_{-1}m_{-2}\cdots m_{-R}}$$

$$= 1 - \frac{1}{m_{-1}m_{-2}\cdots m_{-R}} = 1 - \pi_{-R}. \tag{7.2}$$

As R tends to infinity, π_{-R} tends to zero, and $(1 - \pi_{-R})$ tends to one. The largest expressible mantissa therefore converges to one in any positional system, uniform or mixed. Thus

$$\sum_{i=-1}^{-\infty} (m_i - 1)\pi_i = 1. \tag{7.3}$$

For example,

$$1 = \frac{9}{10} + \frac{9}{100} + \frac{9}{1000} + \cdots \tag{7.4a}$$

$$1 = \frac{1}{2} + \frac{1}{4} + \frac{1}{8} + \cdots \tag{7.4b}$$

$$1 = \frac{1}{2!} + \frac{2}{3!} + \frac{3}{4!} + \cdots \tag{7.4c}$$

Since π_{-R} may be made smaller than any desired positive quantity, an infinite number of infinitely close numbers, in other words, the *continuum* of real numbers, ranging from zero to one, may thus be represented in any positional number system that would be allowed to extend to infinity. For example,

$$e - 2 = \frac{1}{2!} + \frac{1}{3!} + \frac{1}{4!} + \cdots$$

$$\sqrt{2} - 1 = \frac{4}{10} + \frac{1}{100} + \frac{4}{1000} + \frac{2}{10\,000} + \frac{1}{100\,000} + \cdots$$

$$\pi - 3 = \frac{1}{8} + \frac{1}{8.8} + \frac{1}{8.8.17} + \frac{1}{8.8.17.20} + \frac{1}{8.8.17.20.20}$$

$$+ \frac{1}{8.8.17.20.20.11} + \cdots \tag{7.5}$$

Whereas an integer's representation is finite in length, a number's mantissa is *essentially* of infinite length. Whereas each new position to the left of the integer's representation brings with it an increasing finite contribution, a fractional number is the sum of the terms of a convergent infinite series, which terms are also finite, though *evanescent*.

We have seen that rational numbers are always represented by finite or periodic mantissas, when the base is itself periodic. A finite mantissa, such as .25, must be nonetheless regarded as the abbreviation of a periodic representation, in this case, .250000... followed by an infinite number of zeros, or .2499999... followed by an infinite number of 9s. Similarly, $(.1\ 0\ 1\ 1)_2 = (.1\ 0\ 1\ 0\ 1\ 1\ 1\ 1\ldots)_2$ and $(.1\ 1\ 2\ 4)_f = (.1\ 1\ 2\ 3\ 5\ 6\ 7\ 8\ 9\ldots)_f$.

Again, $.\underline{13}$ is the abbreviation of $.131313\ldots$. Though the term $.\underline{13}$ occupies a finite portion of space, it is only a symbol, while the mantissa for which it stands is infinite in length.

To irrational numbers always correspond infinite representations, none of which will ever be fully written out, and computer experts still spend huge amounts of computing power and time calculating π and other transcendental numbers to the nth decimal place, where n is ever larger, knowing perfectly well that no regular pattern will ever emerge.

Contrary to what Kronecker may have declared, namely, that only rational numbers exist in nature, it seems that irrational numbers are the rule, and rational numbers the exception. No instrument ever manufactured, however, will be precise enough to encompass, and no human endeavour will ever require, their overabundant wealth of digits.

So far, we have used such expressions as "tends to infinity," and "tends to zero" loosely, relying on the intuitive connotations carried by these expressions. We need to establish these notions on somewhat firmer ground.

Expressions (7.4a)–(7.4e) and (7.5) are referred to as *infinite series*. They are made up of an infinite number of *evanescent*, or ever-decreasing, terms.

Paraphrasing Anaxagoras (ca. 500–428 B.C.), "There is no smallest among the small, and no largest among the large, but always something still smaller and something still larger." The series under consideration may be extended to any desired length, and made to differ from one by an amount as small as desired.

Using the definition given by French mathematician Jean Le Rond d'Alembert (1717–1783), namely, "A quantity is the limit of another quantity when the second can approach the first more closely than any given quantity, as small as one can suppose," we may state that series (7.4a)–(7.4c) have the number 1 as their *limit*, or that the series *converge* to 1. In the case of equation 7.4c, for example,

$$S_n = \frac{1}{2!} + \frac{2}{3!} + \frac{3}{4!} + \cdots + \frac{n-1}{n!} \qquad \text{converges to 1 as } n \text{ tends to } \infty,$$

$$\text{or} \quad S_n \rightarrow 1 \text{ as } n \rightarrow \infty \quad \text{or} \quad S_n \xrightarrow[n \to \infty]{} 1 \quad \text{or} \quad \lim_{n \to \infty} S_n = 1.$$

The series discussed so far derive from our analysis of positional number systems. When the base is uniform, as in (7.4a) and (7.4b), they are referred to as *geometric series*. Obviously, there are innumerable series of other forms, such as the two remarkable series

$$\frac{\pi^2}{6} = \frac{1}{1^2} + \frac{1}{2^2} + \frac{1}{3^2} + \frac{1}{4^2} + \cdots \tag{7.6}$$

$$\frac{\pi}{4} = \frac{1}{1} - \frac{1}{3} + \frac{1}{5} - \frac{1}{7} + \frac{1}{9} + \cdots \tag{7.7}$$

Series (7.6) was discovered by Leonhard Euler in 1736, and (7.7) was published in 1671 by James Gregory (born in Drumoak, Scotland, in 1638). That series was later rediscovered by Leibniz, and bears his name.

PARADOXES OF INFINITE SERIES

The divergent series are the invention of the devil, and it
is a shame to base on them any demonstration.
(Niels Henrik Abel)

In the discussion of commensurability, a notion dear to Greek hearts, we somewhat hastily stated that the universe of rational numbers was closed with respect to addition and multiplication.

Take addition for example. It is true that adding any two rationals results in a rational. Consequently, adding 1 trillion rationals also results in a rational. But what happens when we add an infinite number of rationals? Do we necessarily obtain a rational? If that were the case, what would you make of the positional representation of irrational numbers? Is it not the sum of an infinity of rational numbers?

Multiplying any two rational numbers also results in a rational number. But is that necessarily the case when we multiply an infinity of rational numbers? Examine the following infinite product, which was discovered in 1650 by the English mathematician and physicist John Wallis (born in Ashford, England, in 1616, author of *Arithmetica infinitorum* and *Opera mathematica*):

$$\frac{\pi}{2} = \frac{2}{1} \times \frac{2}{3} \times \frac{4}{3} \times \frac{4}{5} \times \frac{6}{5} \times \frac{6}{7} \times \cdots . \qquad (7.8)$$

This product illustrates the generation of the transcendental number π by multiplying together an infinity of rational numbers.

Other famous paradoxes have to do with *indeterminate* series, some of which have led astray the greatest among mathematicians. Take the following series:

$$S = 1 - 1 + 1 - 1 + 1 - \cdots. \tag{7.9}$$

It follows from (7.9) that

$$-S = -1 + 1 - 1 + 1 - \cdots \tag{7.10}$$

Let us rearrange series (7.9) and (7.10) as follows:

$$\begin{aligned} S &= 1 - 1 + 1 - 1 + 1 - \cdots, \\ -S &= -1 + 1 - 1 + 1 - \cdots. \end{aligned} \tag{7.11}$$

Subtracting the bottom from the top series, we get

$$S - (-S) = 2S = 1 \quad \text{or} \quad S = 1/2.$$

We may also rewrite the initial series as

$$S = (1 - 1) + (1 - 1) + (1 - 1) + \cdots, \tag{7.12}$$

which adds up to zero; or even as

$$S = 1 + (-1 + 1) + (-1 + 1) + \cdots, \tag{7.13}$$

which adds up to 1. Depending upon which way it is manipulated, the series is equal to 0, 1, or 1/2. The series is *indeterminate*, and cannot be equated to any one of the three values.

What rule authorizes the elimination of the quantities between parentheses in (7.2) and not in (7.13)? Observe that in expression (7.2), the terms inside the parentheses to be subtracted are less than 1, which is not the case in (7.13). But is that generally sufficient?

Quoting Keith Devlin, "Sorting out the distinction between series that can be manipulated and those that cannot, and developing a sound theory of how to handle infinite series, took hundreds of years of effort, and was not completed until late in the nineteenth century."[1]

Consider this other paradoxical series, in which every term, with the exception of the first, is less than 1:

$$S = \frac{1}{1} - \frac{1}{2} + \frac{1}{3} - \frac{1}{4} + \frac{1}{5} - \cdots = \ln 2, \qquad (7.14a)$$

where $\ln 2$ represents the natural (base e) logarithm of 2. Dividing each term of the series by 2, we get

$$S' = \frac{1}{2} - \frac{1}{4} + \frac{1}{6} - \frac{1}{8} + \frac{1}{10} - \cdots = \frac{\ln 2}{2}. \qquad (7.14b)$$

Let us arrange the terms as

$$S' = \qquad \frac{1}{2} \qquad - \frac{1}{4} \qquad + \frac{1}{6} \qquad - \frac{1}{8} + \cdots = \frac{\ln 2}{2},$$

$$S = \frac{1}{1} - \frac{1}{2} + \frac{1}{3} - \frac{1}{4} + \frac{1}{5} - \frac{1}{6} + \frac{1}{7} - \frac{1}{8} + \cdots = \ln 2.$$

We get

$$S + S' = \frac{1}{1} + \frac{1}{3} - \frac{1}{2} + \frac{1}{5} + \frac{1}{7} - \frac{1}{4} + \cdots = \frac{3\ln 2}{2}. \qquad (7.14c)$$

Series $S + S'$ is a mere *rearrangement* of the terms of S, leading to the absurd conclusion that $\ln 2 = (3\ln 2)/2$.

A series whose convergence is due to the alternating signs of its terms is said to be *conditionally convergent*, whereas a series that converges irrespective of the signs is said to be *absolutely convergent*. It has been established that rearrangement of a series' terms is authorized only if it belongs to the latter category.

---

[1] Keith Devlin, *Mathematics, the Science of Patterns* (New York: Scientific American Library, 1994).

Another famous series that has baffled generations of mathematicians, is known as the *harmonic series*:

$$h = \frac{1}{1} + \frac{1}{2} + \frac{1}{3} + \frac{1}{4} + \frac{1}{5} + \cdots. \qquad (7.15)$$

The fourteenth-century French scholar Nicolas Oresme (born ca. 1360), author of *Tractatus de figuratione potentarium et ensurarum*, a truly visionary precursor of calculus, proved that the harmonic series diverges, in other words, that it tends to an infinite value as the number of terms tends to infinity.

His proof is obtained by rearranging the series as follows:

$$h = 1 + \frac{1}{2} + \left(\frac{1}{3} + \frac{1}{4}\right) + \left(\frac{1}{5} + \frac{1}{6} + \frac{1}{7} + \frac{1}{8}\right)$$
$$+ \left(\frac{1}{9} + \cdots + \frac{1}{16}\right) + \cdots$$

Series h' is now constructed by replacing every denominator within any pair of parentheses in h by the largest denominator within the parentheses. Thus,

$$h' = 1 + \frac{1}{2} + \left(\frac{1}{4} + \frac{1}{4}\right) + \left(\frac{1}{8} + \frac{1}{8} + \frac{1}{8} + \frac{1}{8}\right)$$
$$+ \left(\frac{1}{16} + \cdots + \frac{1}{16}\right) + \cdots$$

Clearly, the sum within any given parentheses in h' is smaller than the corresponding sum in h. It follows that $h > h'$. Since the sums between parentheses of h' are all equal to 1/2, it is clear that h' diverges. Consequently h also diverges, despite the fact that its terms are evanescent. The property of evanescence of the successive terms of a series does not, therefore, constitute a sufficient condition for convergence.

Harmonic series h owes its name to the kinship between its terms and the harmonic vibrations of a stretched string. Its behavior has been studied at length, and the reader is advised not to attempt to demonstrate its divergence using a home computer. According to Eli Maor, it would take a fairly fast computer the age of the universe squared[2], to allow the series to approach 100. That is very slow indeed. The great Euler has nonetheless shown that as the number N of terms tends to infinity,

$$h = \frac{1}{1} + \frac{1}{2} + \frac{1}{3} + \frac{1}{4} + \frac{1}{5} + \cdots + \frac{1}{N} \xrightarrow[N \to \infty]{} \ln N + \gamma, \quad (7.16)$$

where γ is known as *Euler's constant*, and is approximately equal to .57722.

The difficult problem of convergence will not be addressed here beyond the preceding discussion, whose aim was merely to illustrate the potential paradoxes concealed by infinite series.

Further Paradoxes of Infinity

> Eternity is very boring,
> specially near the end.
> (Quote attributed to Woody Allen)

Let us return to Equation (7.3).

$$\sum_{i=-1}^{-\infty} (m_i - 1)\pi_i = 1.$$

That statement is true whether or not m_i is an integer, provided it is a positive number larger than 1. Let us examine what happens with the following choice of "base" b :

$$b = \left(\cdot\frac{2}{1}, \frac{3}{2}, \frac{4}{3}, \frac{5}{4}, \cdots \right).$$

We obtain

$$\frac{\frac{2}{1} - 1}{\frac{2}{1}} + \frac{\frac{3}{2} - 1}{\frac{2}{1} \times \frac{3}{2}} + \frac{\frac{4}{3} - 1}{\frac{2}{1} \times \frac{3}{2} \times \frac{4}{3}} + \cdots = 1.$$

[2] Eli Maor, *To Infinity and Beyond* (Princeton, N.J.: Princeton University Press, 1991).

Thus,

$$\frac{1/1}{2} + \frac{1/2}{3} + \frac{1/3}{4} + \frac{1/4}{5} + \cdots = 1, \qquad (7.17)$$

which may be also written as

$$\frac{1}{1 \times 2} + \frac{1}{2 \times 3} + \frac{1}{3 \times 4} + \frac{1}{4 \times 5} + \cdots = 1. \qquad (7.18)$$

Comparing (7.18) with the harmonic series, we may write

$$\frac{1/a_1}{2} + \frac{1/a_2}{3} + \frac{1/a_3}{4} + \cdots + \frac{1/a_n}{n+1} \xrightarrow[n \to \infty]{} 1 \quad \text{when } a_n = n,$$

$$\xrightarrow[n \to \infty]{} \infty \quad \text{when } a_n = 1.$$

$$(7.19)$$

Figure 7.1 provides a geometric interpretation of (7.18) that was drawn for a small number of triangles and must be imagined to extend to infinity. The sum of the areas under the triangles, which we may refer to as *harmonic triangles*, is equal to

$$\frac{1}{2}\left(\frac{1}{1 \times 2} + \frac{1}{2 \times 3} + \frac{1}{3 \times 4} + \frac{1}{4 \times 5} + \cdots \right) = \frac{1}{2}, \quad (7.20a)$$

and the length of the horizontal axis upon which the triangles are raised is equal to

$$\frac{1}{2} + \frac{1}{4} + \frac{1}{6} + \frac{1}{8} + \cdots = \frac{h}{2} = \infty.$$

It is clear from the figure that the broken curve is longer than the sum of the horizontal bases, which is infinite. We are therefore in the presence of a paradoxical geometric figure whose area is finite, and whose perimeter is infinite.

That is also the case of the hyperbola, among other figures, including the figure known as the *Koch curve* or *snowflake*. The Koch curve, in addition to having a finite area and infinite perimeter, is such that the distance between any two of its points, measured along the perimeter, is also infinite.

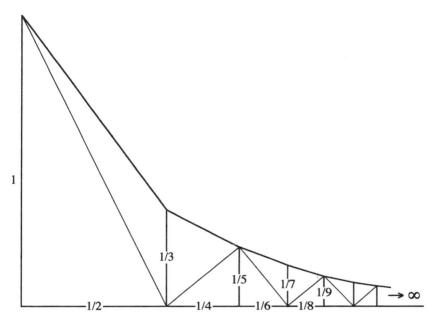

Figure 7.1. The "harmonic triangles."

It can also be easily established that

$$\frac{1}{2} = \frac{1}{2 \times 3} + \frac{1}{3 \times 4} + \frac{1}{4 \times 5} + \cdots,$$

$$\frac{1}{3} = \frac{1}{3 \times 4} + \frac{1}{4 \times 5} + \frac{1}{5 \times 6} + \cdots,$$

and in general,

$$\frac{1}{n} = \frac{1}{n \times (n+1)} + \frac{1}{(n+1) \times (n+2)}$$
$$+ \frac{1}{(n+2) \times (n+3)} + \cdots \qquad (7.20\text{b})$$

The latter result may be easily proven by induction over n.

You Are Always Welcome at the Hilbert Hotel

The infinite! No other question has ever moved so
profoundly the spirit of man; no other idea has so fruitfully
stimulated his intellect; yet no other concept stands in
greater need of clarification than that of the infinite.
(David Hilbert)

In manipulating the series (7.10), we have implicitly accepted that
if the initial element of an infinite collection of objects is removed,
the resulting collection may be matched, or paired object for object,
with the initial collection. In manipulating series (7.14a)–(7.14c), we
have similarly accepted that since the removal of the first object of
an infinite collection leaves an equally infinite collection, removing
the third, fifth, seventh objects, etc., has the same effect. We concluded that removing every odd-rank object still leaves an infinity of
objects that can still be paired one for one with those of the initial
collection.

David Hilbert (1862–1943), a turn-of-the-century German mathematician, told the following fictitious anecdote to illustrate the paradox underlying that observation: The Hilbert Hotel has an infinity of
rooms, all of which are full, and a traveling mathematician wishes to
check in. The clerk apologizes, declaring that no vacancies are available. Thereupon, the mathematician suggests that the problem may
be easily solved by inviting the tenant of room 1 to move to room 2,
whose occupant would be asked to move to room 3, and so on. The
number of rooms being infinite, no customer will fall off the edge as
the process ripples through the entire hotel, from one room to the
next.

In *Dialogues of Two New Sciences*, Galileo contemplates infinity
with great awe: "Infinities and indivisibles transcend our finite understanding, the former for their magnitude, the latter for their smallness.
Imagine what they are when combined." In one of the imaginary dialogues between Simplicio and himself, alias Salvati, Galileo observes
that square numbers, which are interspersed among the natural numbers, become more and more sparsely distributed as the numbers
get larger. There should therefore be less squares than natural num-

bers. But each square number is none other than a natural number multiplied by itself, leading to the conflicting conclusion that squares and natural numbers must be equal in number. Galileo dismissed the dilemma by concluding that the notions of equality and inequality were not relevant when dealing with infinity.

It is impossible to discuss the infinite without giving credit to Bernhard Bolzano (1781–1848), the Czech mathematician, logician, and philosopher of Italian origin who is regarded as the true founder of set theory. In his *Paradoxien des unendlichen* (Paradoxes of the infinite), published in 1851, the word "set" appears for the first time in mathematical language. His work was not given the attention it deserved at the time, though it contained the germ of the later work of Dedekind and Cantor, and inspired both mathematicians. He defined *reflexivity* as the property of infinite sets of being able to be put into one-to-one correspondence with some of their proper subsets, and, indeed, as the very definition of infinite sets.

Zeno's Paradoxes

Greek scholars are privileged men;
few of them know Greek,
and most of them know nothing else.
(George Bernard Shaw)

Every high school student has been exposed at some time or another to the tale of Achilles and the tortoise, one of the four paradoxes of Zeno the Eleate, a student of Parmenides who lived in the fifth century B.C. Achilles runs twice as fast as the tortoise, and challenges it to a race, granting it a head start of, say, 10 meters. By the time the swift Achilles covers that initial distance, the slow but unrelenting tortoise (an attribute glorified by Jean de la Fontaine in his memorable fable of the race between the tortoise and the hare), has covered 5 meters. By the time Achilles covers that distance, the tortoise has covered 2.5 meters, and so on. After all eternity has elapsed, the tortoise remains half of some residual distance ahead, and Achilles never catches up. Using the same argument, Zeno came to the conclusion that an arrow's movement was but an illusion: In order to reach its destination,

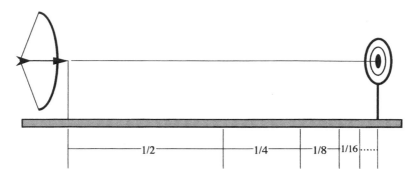

Figure 7.2. Zeno's paradox.

the arrow first has to cover half the distance between the archer and the target, then half of the remaining distance, then half of that half, and so on, indefinitely (Figure 7.2).

Since every paradox had to be dismissed, Zeno declared that motion simply did not exist, and founded his philosophy on that radical—and useless—postulate. Modern-day students scoff at the paradox, and accept that the infinite series $1/2 + 1/4 + 1/8 + \ldots$ converges to 1. In their minds, Zeno's tribulations no longer deserve another thought, unaware as they are of the yet unresolved and perhaps insurmountable paradoxes concealed in three little suspension dots.

Horror Infiniti?

> There was more imagination in the head
> of Archimedes than in that of Homer.
> (Voltaire)

Aristotle wrote, "Infinity is not perfection. It is privation, the absence of a limit." *Horror infiniti*, an alleged aversion harbored by the ancient Greeks towards infinity, did not prevent them from putting the notion of infinity to use with extraordinary acumen. Take the constant

π for instance. Whereas the ancient Egyptians and Mesopotamians were content with approximate values that they could put to practical use in architecture and other earthly endeavours, the Greeks were more concerned with the method, and with the irrefutability of its logical grounding. The Egyptian method consisted of inscribing a circle in a square that was subsequently transformed into an octagon by shaving off its corners (Figure 7.3a). With a square of side $D = 9$ units, the octagon's area was $7D^2/9 = 63$. The Egyptians posited that the circle's area, whose diameter was also 9 units, was $8^2 = 64$, in other words, that

$$\pi \left(\frac{9}{2}\right)^2 \approx 8^2,$$

that is,

$$\pi \approx \left(\frac{8 \times 2}{9}\right)^2 \approx 3.1605.$$

The Mesopotamians posited that the ratio of the perimeter of an inscribed hexagon to a circle's circumference was $(.57 \ 36)_{60} = 57/60 + 36/60^2 = 0.96$, thus setting the value of π at 3.125.

In attempting to square the circle, Antiphon, a Greek philosopher (born ca. 500 B.C.), studied the Egyptian and Mesopotamian approaches, both of which were predicated upon inscribing a polygon in a circle. He is credited with inventing the method of *exhaustion*, which consists of inscribing a regular polygon in a circle, and then doubling the number of its sides again and again until the circle is hopefully *exhausted*, allowing one to equate its circumference with the polygon's perimeter.

In Book XII of his *Elements*, Euclid enunciated, "Circles are to one another as the squares on the diameters," thereby affirming that a proportionality constant existed, which applied to all circles, regardless of the diameter. That same constant also governed the relationship between a circle's diameter and its circumference.

Most certainly aware that the exhaustion process might never come to an end, he showed that the difference between the polygon's perimeter and the circle's circumference could be made *smaller*

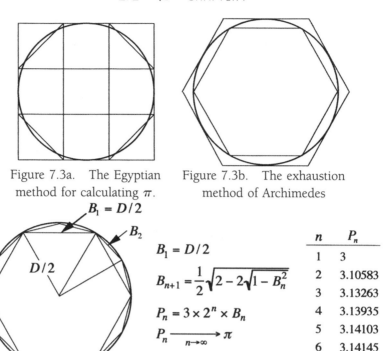

Figure 7.3a. The Egyptian method for calculating π.

Figure 7.3b. The exhaustion method of Archimedes

$$B_1 = D/2$$

$$B_{n+1} = \frac{1}{2}\sqrt{2 - 2\sqrt{1 - B_n^2}}$$

$$P_n = 3 \times 2^n \times B_n$$

$$P_n \xrightarrow[n \to \infty]{} \pi$$

n	P_n
1	3
2	3.10583
3	3.13263
4	3.13935
5	3.14103
6	3.14145
	$\cdots$

Figure 7.3c. Convergence to π. *Source*: Rudy Rucker, *Infinity and the Mind* (New York: Bantam, 1983).

than any arbitrarily chosen quantity. In so doing, Euclid implicitly established the legitimacy of limiting processes, two millenia before the French mathematiciam Cauchy, who, together with the German mathematician Karl Weierstrass, afforded a secure foundation for the calculus invented by Newton and Leibniz.

Archimedes, who studied at the University of Alexandria, used that method to calculate π within limits as narrow as desired. That consisted of inscribing and circumscribing polygons with an ever-increasing number of sides, thus approaching the number π from above and below (Figure 7.3b). The construction of Figure 7.3c yields a simple algorithm offered by Rudy Rucker for the calculations of π's lower bound with any desired level of accuracy. With a 96-sided polygon, Archimedes arrived at the remarkable statement

$$3\frac{10}{71} < \pi < 3\frac{1}{7}. \qquad (7.21a)$$

Calculating π was not the only instance of flirting with infinity by the ancient Greeks. As one descends the Ladder of Theodorus of Cyrene, the successive rational numbers obtained are alternately less than and greater than $\sqrt{2}$, but inexorably closer and closer to what we today call a limit. Theodorus was thus able to also make remarkable statements regarding the square root of 2, as well as the square roots of several other integers, such as

$$\frac{239}{169} < \sqrt{2} < \frac{99}{70}. \qquad (7.21b)$$

These figures are obtained as early as the ladder's sixth rung.

Exhaustion processes led the Greeks, willy-nilly, to grapple with infinite processes. The notion of physical nonfinitude was culturally abhorrent to the early Greeks, and these processes could obviously not be accomplished within a finite time frame. The Greeks could not conceive of an infinite number of numbers, but perhaps suspected the existence of a very large last number. In the words of Tobias Dantzig, "It is a plausible hypothesis that the early conception of infinity was not the uncountable, but the yet-uncounted." That view was consistent with atomism, a prophetic vision expounded by Democritus (b. ca. 470 B.C.), among others, to whom matter consisted of a very large number of very small, indivisible grains, or atoms. The subdivision of matter could not be carried out forever.

Perhaps one of the most profound instances of resorting to infinite processes is concealed in the truly marvelous definition of the equality of irrational numbers offered by Eudoxus of Cnidus. His definition was stated in rather simple terms, and may be restated in yet simpler modern terms: two irrational numbers are equal if no rational number can be found that is smaller than the first and greater than the second. Simple indeed. But how do you ascertain that there is no such number, unless you embark on an infinite search? The statement itself may indeed be short and finite, but its implementation is infinite.

POTENTIAL VERSUS ACTUAL INFINITY

We can assert with certitude that the universe is
all Center, or that the center of the universe is
everywhere and the Circumference nowhere.
(Giordano Bruno)[3]

Eudoxus's definition is grounded in infinite processes. So is Dedekind's, whose *Schnitt* is only meaningful within the realm of infinite sets of rational numbers. Irrational numbers require a framework where infinity is accepted not only as some kind of limiting process but as an entity that exists in actuality.

We have seen that multiplying an extremely large number of rational numbers yields a rational number, no matter how large. It is only when that number of numbers is *actually* infinite that the product may be truly irrational. That observation may shed some light on the heated debate that stemmed from Aristotle's definition of *potential infinity*, as opposed to *actual infinity*. Aristotle postulated that actual infinity does not exist; infinity only constitutes some kind of ideal toward which one may only tend. In his *Physics*, he wrote: "The infinite has a potential existence . . . There will not be an actual infinite." In other words, there are only infinite processes, not infinite objects. The opposite view, which eventually triumphed with Georg Cantor, is that infinite sets exist as such in *actuality*, and may be treated "all at once," "d'emblée," "le tout ensemble." There exist finite sets as well as infinite sets of objects. These sets are distinct in nature, and each may be considered in its totality, with its own distinctive properties, which properties make it qualitatively different from the others.

For centuries, the belief that Aristotle had finally solved Zeno's paradoxes was deeply entrenched in Western culture. In his view, the distance between Achilles and the tortoise is finite in actuality; the athlete eventually overtakes the tortoise in an observably finite amount

---

[3]Giordano Bruno (1548–1600), an Italian philosopher born in Nola, was expelled from the order of the Dominicans after being accused of heresy, imprisoned by the Calvinists in Geneva, excommunicated by the Lutherans in Germany, and finally burned alive by the inquisition after seven years of imprisonment, as "an impenitent heretic, stubborn and obstinate." *Della causa, principio ed uno* (On the cause, the principle and unity), quoted by Jorge Luis Borges in "The Fearful Sphere of Pascal."

of time. The distance between the runners may nonetheless be subdivided into a never-ending sequence of smaller and smaller segments. Nothing stands in the way of imagining such a process, which may not be completed within a finite time frame. It spans a kind of infinity that exists only in potentiality, not in actuality. According to Aristotle, infinity may not be traversed as one traverses a finite stretch of space, and it lacks a limit. Without limit, it may not be circumscribed, and may not, therefore, be examined as an entity. If infinity were an entity, its parts would also be infinite, and we would be dealing with a hierarchy of infinities, some of which might be larger than others, or with a system where the part was as large as the whole, in total contradiction to the last postulate in Euclid's *Elements*. Neither alternative was acceptable to Aristotle, and it took the genius of Cantor to embrace actual infinity, declaring that the part can be as large as the whole, and recognizing the existence of a hierarchy of infinities.

CANTOR

> An infinite set is one that can be put into a one-to-one correspondence with a proper subset of itself.
> (R. Dedekind)[4]

Let us for a moment return to the number lattice. An infinity of lines may be shot out of the origin without ever touching a node! That is not the whole story: Georg Cantor (born in Saint Petersburg on March 3, 1845) showed that the infinity of lines that never encounter a node is *larger* than the infinity of those lines that do! But what exactly did he mean by larger?

To answer that question, let us begin at node (1,1) of Figure (7.4a), and devise a geometric scheme for exploring the prime nodes in succession. Cantor's scheme yields the following *enumeration*:

$$1/1, 2/1, -2/1, -1/1, -1/2, -3/2, -3/1, 3/1, 3/2,$$
$$1/2, 1/3, 2/3, 4/3, 4/1, -4/1, -4/3 \ldots.$$

---

[4]*Was sind und was sollen die Zahlen* (1888).

Georg Cantor (1945–1918). Courtesy Deutsches Museum, Munich.

The enumeration process is considerably simplified if we observe that the distribution of prime nodes exhibits three different symmetries:

1. Within the first <u>octant</u>, the prime node pattern on any given abscissa b is symmetrical about the ordinate's midpoint. That is because if the numbers b, a are relatively prime, so are the numbers $b, b - a$. Euler's number $\phi(b)$ is always an even number.
2. The prime node pattern in the first <u>quadrant</u> is symmetrical about the unit line, meaning that if (a, b) is a prime node, so is (b, a).
3. The prime node pattern is also symmetrical about the <u>y axis</u>.

To integer pair 1, 1 corresponds one and only one prime node, namely, $(1, 1)$, and to integer pair 1, 2 correspond prime nodes $(1, 2)$ and $(2, 1)$. To any integer pair a, b such that

$$2a < b \quad \text{and} \quad a \perp b \tag{7.22}$$

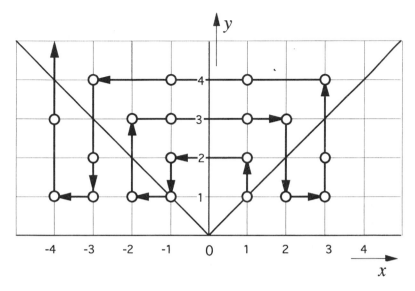

Figure 7.4a. Cantor's enumeration.

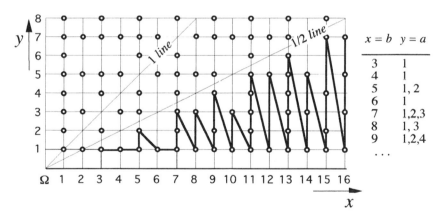

Figure 7.4b. Alternative prime node enumeration.

correspond four *distinct* positive prime nodes, namely,

$$(a, b), (b - a, b), (b, a), (b, b - a). \qquad (7.23)$$

Conversely, any prime node (a, b) other than $(1, 1)$, $(1, 2)$, or $(2, 1)$ satisfies (7.22) and (7.23).

Based on these observations, the enumeration scheme in Figure 7.4b is derived from that utilized in Figure 6.3. It consists of listing for every denominator $b > 2$ those integers a such that $a \perp b$, $2a < b$, with the knowledge that to each integer a corresponds a set of four prime nodes, as in (7.23).

Nodes $(1, 1)$, $(1, 2)$, and $(2, 1)$ need to be added to the above enumeration. The prime node pattern is symmetrical about the vertical axis. To every accounted-for positive rational also corresponds a symmetric node about the vertical axis, defining a negative rational.

It must be made clear that the process of denumeration is totally distinct from that of *ordering*. No attempt at ordering rationals will ever succeed, as it is always possible to squeeze an infinity of rationals between any two given rationals. In the words of Tobias Dantzig, "We have obtained succession at the expense of continuity."

The *enumerability* of rational numbers led Cantor to elaborate a revolutionary theory of the infinite, which to this day has its ardent supporters, as well as its no less ardent detractors. He introduced the notion of infinities of different magnitudes, or *powers*, where an infinite set is said to be of the same power as that of the infinite set of natural numbers if a one-to-one correspondence can be established between their elements; in other words, if the set is *denumerable*, or *countable*. According to that definition, the denumerable set of rational numbers is of the same power as that of the set of natural numbers— counterintuitively, since common sense would seem to dictate that each rational number requires two natural numbers, suggesting that there might be twice as many rationals as there are integers.

The Power of the Continuum

Using a clever trick, that of an equation's *height*, Cantor showed that the aggregate of algebraic numbers is also denumerable. With Cantor's definition, the aggregate of integers is of the same power as that of the integers, rationals, and algebraic numbers, all taken together. The criterion of denumerability authorizes one to equate a part to the whole, thus alleviating the ambiguities encountered in the dialogue between Galileo, alias Salvati, and Simplicio. In the course of his initial attempt to prove that the continuum of real numbers is denumerable, Cantor discovered that it is not, and contended that there are incomparably more transcendental than algebraic numbers. He showed that whereas the power of the algebraic domain is that of the natural numbers, the power of of the transcendental number

TABLE 7.1
Cantor's diagonalization.

$$r_0 = . \; \delta^{r_0}_{-1}, \; \delta^{r_0}_{-2}, \; \delta^{r_0}_{-3}, \; \delta^{r_0}_{-4}, \; \ldots$$

$$r_1 = . \; \delta^{r_1}_{-1}, \; \delta^{r_1}_{-2}, \; \delta^{r_1}_{-3}, \; \delta^{r_1}_{-4}, \; \ldots$$

$$r_2 = . \; \delta^{r_2}_{-1}, \; \delta^{r_2}_{-2}, \; \delta^{r_2}_{-3}, \; \delta^{r_2}_{-4}, \; \ldots$$

$$r_3 = . \; \delta^{r_3}_{-1}, \; \delta^{r_3}_{-2}, \; \delta^{r_3}_{-3}, \; \delta^{r_3}_{-4}, \; \ldots$$

. . .

aggregate is that of the real number continuum, which he proved to be nondenumerable. His proof is based on another clever trick, that of diagonalization, and proceeds by reductio ad absurdum, as follows: If the aggregate R of real numbers were denumerable, we could list *all* its members $r_0, r_1, r_2, \ldots$ in succession. Imagine that the resulting hypothetical exhaustive list of fractional numbers is written in some kind of positional number system, as shown in Table 7.1. Imagine now that the number a is constructed following the rule

$$\delta^a_{-1} \neq \delta^{r_0}_{-1}, \quad \delta^a_{-2} \neq \delta^{r_1}_{-2}, \quad \delta^a_{-3} \neq \delta^{r_2}_{-3}, \quad \delta^a_{-4} \neq \delta^{r_3}_{-4}, \ldots$$

Clearly, a is different from every number present in the list. That list therefore turns out to be nonexhaustive, contradicting the premises.

Geometrical Metaphors

Points
Have no parts or joints.
How can they combine
to form a line?
(J. A. Lindon)

In the geometrical metaphor, any finite line segment contains an infinity of points, whose power is that of the continuum. That number is the same regardless of the segment's length, whether the line under consideration is finite or infinite, as shown in Figure 7.5, where the straight rays shot from pole 0 correspond one to one with the points on the datum line.

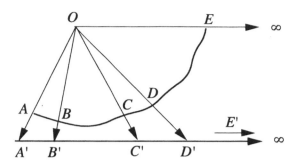

Figure 7.5. Projection of a finite line segment upon an infinite ray.

We may now legitimately ask ourselves if the number of points in the plane is a greater cardinal than that of the line. Cantor spent three years trying to prove that it was, until he discovered in 1877 that it was not.

Consider a square whose one-unit sides are parallel to the Cartesian coordinate axes, and whose lower left corner coincides with the origin (Figure 7.6). Consider now a point within the square whose abscissa x is equal to the mantissa of $\sqrt{2}$, and whose ordinate y is equal to the mantissa of the Golden Ratio. In other words, $x = .414213562\ldots$, and $y = .618033989\ldots$.

Cantor invented a scheme for generating a number x' from any two numbers x and y by interspersing the digits of y within those of x, according to some fixed unchanging rule. We may for instance decide

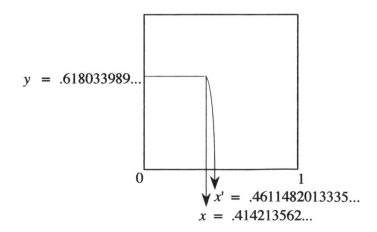

Figure 7.6. Cantor's construction.

that $x' = .46\underline{1}14\underline{8}20\underline{1}33\underline{3}59\underline{6}82\underline{9}\ldots$. In that scheme, the under-
lined digits belong to x, and the remaining digits belong to y. If x' is
regarded as the abscissa of a point whose ordinate is zero, a one-to-one
correspondence may be performed between the points on the plane
and those on the x-axis, thus establishing that they possess the same
power.

Cantor's trick may be easily extended to the n-dimensional hyper-
cube by interlacing the digits of a point's n coordinates.

Transfinite Cardinal Numbers

If the process of matching two collections (or aggregates), element
for element, exhausts one collection and leaves some unaccounted-for
elements in the other, the latter collection is said to be of a greater
power than the first. When dealing with *finite* collections of objects,
collections of the same power may be assigned some cardinal number.
If two collections are not of the same power, a larger cardinal number
is assigned to the greater of the two. Cantor's leap forward consisted
of assigning *transfinite cardinals* to the powers of infinite collections,
and inventing a corresponding transfinite arithmetic, where transfinite
cardinals are treated as bona fide numbers, comparable to other aggregates examined so far: the natural numbers, the rational numbers, the
algebraic irrationals, and the transcendental numbers. Cantor's ordered
sequence of transfinite cardinals begins with the power $\aleph_0$, or *aleph
null*, for the denumerable aggregates. There is no smaller transfinite.
The power of the continuum is $\aleph_1$, or *aleph 1*. There are as many
transcendental numbers as there are numbers in the continuum, and
the latter also contains integers, rational numbers, and algebraic numbers. (By now, such a statement should not be found shocking, accustomed as we have become to accepting that the part may be as large as
the whole.) Cantor postulated that there exists no intermediate power
between that of the natural numbers and that of the continuum, and
to this day, no aggregate has been found whose power is higher than
$\aleph_0$ and lower than $\aleph_1$.

Cantor Dust

A well-known figure introduced by Cantor to illustrate the paradoxes of the infinite is known as the *Cantor Dust*. That figure is constructed as follows: Begin with the straight line at stage 0 in Figure 7.7.

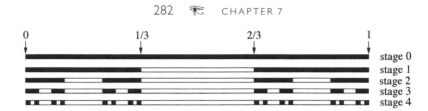

Figure 7.7. The Cantor Dust.

That line has unit length. It includes origin 0, and extends infinitely close to the 1 mark, without containing it. Remove its middle third part, which consits of the 1/3 mark and extends infinitely close to the 2/3 mark, without containing it. Stage 1 is obtained. Proceed with the following stages shown in the figure, each time removing the middle third of each remaining line together with its origin, but without its end point. At stage n, the gaps are $2^n - 1$ in number, and so are their untouched end points: 2/3 at stage 1, then 2/3, 2/9, 8/9 at stage 2, then 2/3, 2/9, 8/9, 2/27, 8/27, 20/27, 26/27 at stage 3, and so on. It therefore seems that as the line's points are gradually depleted, more and more end points emerge. As the process is extended to infinity, we are left with infinitely many end points, separated by infinitely many gaps, each containing infinitely many points. Stranger still, the remaining *dust* is of the same power as the continuum; it is not denumerable. To prove that point, we shall resort to the ternary, or triadic, positional number system.

Let us turn back to Figure 2.10, which show a triadic yardstick that can measure the distances included between 0 and 1 (actually between 0 and as close to 1 as we wish) in increments of $1/3^n$, where n is the number of stages. Applying the yardstick to the Cantor Dust, we discover that the dust does not contain any point whose coordinate's mantissa contains the digit 1. The remaining mantissas consist of 0s and 2s in every possible infinite-length configuration. If we replace the digit 2 by 1, and regard the new configurations as base 2 mantissas, we discover that every number between 0 and 1 is represented, in other words, the *continuum*. The remaining dust thus contains as many points as the original line!

Beyond Aleph I

Consider a finite *set* S (Figure 7.8) containing three elements, namely, A, B, C. Using these elements, one may assemble $2^3 = 8$ dif-

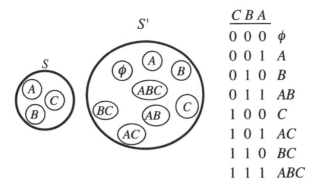

C B A	
0 0 0	φ
0 0 1	A
0 1 0	B
0 1 1	AB
1 0 0	C
1 0 1	AC
1 1 0	BC
1 1 1	ABC

Figure 7.8. Finite set S and its subsets.

ferent *subsets*, namely, *singles* (*A*, *B*, *C*), pairs (*AB*, *AC*, *BC*), the 3-*tuple* (*ABC*), and the *empty set*, denoted *φ*). The figure also highlights the correspondence between these subsets and their binary representation.

Let S' denote the set of subsets of S. If one attempts to establish a one-to-one correspondence between the elements of the two sets, five elements of S will remain unaccounted for. The "power" of S' is obviously greater than that of S. Is the same true of infinite sets? In other words does the set of all subsets of the real number continuum contain more elements than aleph 1 contains? In proving that it does, Cantor offered the world one of the most brilliant proofs in set theory. The proof proceeds as follows.

Consider the infinite set E shown in Figure 7.9, all of whose subsets are formed, thoroughly exhausting every possible combination of its elements in *singles*, *pairs*, 3-*tuples*, etc. Let E' be the aggregate of all these subsets. Reasoning by reductio ad absurdum, let us posit that *all* the elements of E have been paired one for one with those of E', following some kind of random but exhaustive lottery.

At the end of the pairing process, let us examine each element of E together with the subset within E' with which it was paired. We color that element white if it is contained within the subset with which it was paired, and black if it is not. For example, if the process matches A with BCFG, then A is colored black, whereas if B is matched with BFHK, it is colored white. Remember that whereas each element A, B, C, ... belongs to an infinity of subsets, it is assumed to be paired with one and only one subset of aggregate E'. We shall now go one step further, and look within E' for the subset B that contains *all* the black elements, and *only* these. That subset must exist, since

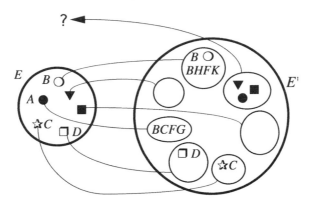

Figure 7.9. Cantor's impossible mapping.

we made sure that the subset formation scheme would leave no subset unformed. We eventually find it in a corner of E'. According to our pairing hypothesis, we are confident that B has been paired with some element of E. As we consult the pairing table, we make the following disconcerting discovery:

1. If B is paired with a white element, it is a faulty pairing, since a white element can be paired only with a subset containing it, and B contains *only* black elements.
2. If B is paired with a black element, it is also a faulty pairing, since a black element can be paired only with a subset that does not contain it, and B contains *all* the black elements.

The conclusion is that no infinite set may be put into one-to-one correspondence with all of its subsets. The power of the aggregate of a set's subsets is greater than that of the set itself. The power of the set of all integers, rationals, and algebraic numbers is aleph null. That of the continuum is aleph 1, and that of the aggregate of the continuum's subsets is aleph 2.

According to Cantor, then, there exists a hierarchy of infinities, corresponding to aggregates of aggregates of aggregates . . . (The words "set" and "aggregate" are used interchangeably in here.)

A famous paradox, referred to as "Russell's paradox," arises, however, when considering the "set of all sets." We shall not dwell upon that paradox, which was simply dismissed by Cantor with the statement that there is no such thing as a set of all sets, and consequently no last transfinite.

Does aleph 2 correspond to anything we know in everyday mathematics? The answer is suggested by considering that it is the set of all permutations of a line's points. It is therefore the set of all curves that may be drawn on a plane, in other words, the set of all functions of a variable. What are aleph 3, 4, . . .? No one knows. In a fairly ancient *Scientific American* article by the great Martin Gardner, the *aggregate* of whose columns constitutes a true fountainhead, Gardner points out that aleph 2 is the number of all possible laws of science, and appropriately quotes George Gamow: "We find ourselves here in a position exactly opposite to that of . . . the Hottentot who had many sons but could not count beyond three!"

In the preceding, we barely scratched the surface, in the hope of providing the reader with an elementary introduction to this vast and difficult subject, which belongs to philosophy as much as to mathematics, and whose protagonists base their views on their own intuitive grasp of infinity, leaving the matter open to opinion and subjective interpretation. Kant himself was an "actualist" early in life, and then became a "potentialist," only to fall back on his earlier beliefs later in life. Over the centuries, infinity has been at the center of heated battles, some of which are being waged to this day.

Finally, I shall quote A. W. Moore, tutorial fellow in philosophy at Oxford: "I would urge mathematicians and other scientists to use more caution than usual when assessing how Cantor's results bear on traditional conceptions of infinity. The truly infinite, it seems, remains well beyond our grasp."[5]

POSTSCRIPT: THE BALANCE IS IMPROBABLE, BUT THE NIGHT SKY IS BLACK

According to Cantor, there are *infinitely many more* irrational than rational numbers in the real number continuum. With that in mind, imagine a perfectly balanced wheel of fortune, with every number from 0 to 1 (or rather, infinitely close to 1), inscribed upon its periphery. The probability of the wheel's coming to rest upon a rational number is zero. No matter how skilled he may be, a craftsman will therefore

---

[5]A. W. Moore, "A Brief History of Infinity," *Scientific American*, April 1995.

never be able to place an infinitely sharp fulcrum at the exact center of a straight lever of some given fixed length: he will always fall upon some infinitely close irrational number. Assuming the lever's total length to be 1, the fulcrum will fall upon a point whose distance is a from one extremity, and $(1 - a)$ from the other. It is easily seen that $a/(1 - a)$ is irrational if a is irrational. Thus, even if one uses ideally equal little beads, the scales will never achieve equilibrium, no matter how many beads are used on either side of the scales.

According to Cantor, then, the weighing scales, which have been in use several millennia, are simply *Improbable*! Fortunately, we live in a world where there is no perfectly sharp fulcrum, and where exact centers and ideally equal little beads exist only in our imagination.

Return for a moment to the previous chapter, and imagine that the number lattice represents the Universe, with every node, prime or not, materialized by a star. Standing at point Ω and looking around you in every direction in the thick of night, you will observe that the heavens are populated with an infinitely large number of shining stars, against the backdrop of an infinitely large black sky . . .

☥ Subject Index ☥

abacus, 32, 42; Asian, 50
Abbassid period, 44
absolutely convergent series, 263
Abu Simbel, 20
Academy, 178, 182
actual infinity, 274–75
adder, parallel, constructing, 78–79
addition, 72–74; associative law of, 155; associativity of, 157; commutative law of, 155; commutativity of, 157; Egyptian, 24
additive identity, 157
additive identity law, 155
additive inverse, 157
additive inverse law, 155
additive systems, xii, 10
affine transformations, 224–26
alchemy, 4
aleph 1, 281; beyond, 282–85
aleph 2, 284, 285
aleph null, 281
Alexandria, 43
Alexandrian Library, 38
algebraic numbers, 98
algoristic textbooks, 48
algorithm, 45
Al-Hawi f'il hissab, 76
al-jabr, 44
Al-jabr W'Al-Moqabala, 43, 44
al-Moqabala, 37
alogon, 163
alphabet, cuneiform, 17
Antichrist, 4
arabic numerals, 46
Arabs, 43–45
ARCL test, 118
Argand diagram, 199
argument, 84
arithmetic: Egyptian, 24–28; floating point, 52; fundamental theorem of, 107–9
ars abaci, 46
Asian abacus, 50

associative laws of addition and multiplication, 155
associativity: of addition, 157; of multiplication, 157
astrology, 4
astronomy, 4
atomism, 273
automata, 246; cleavages and, 246–48
axiomatic method, 158–60

Babylonian educational game, imaginary, 32–33
Baghdad, 43
base, 64
base b cleavage sequence for π, 243
base radices, 103
base ranges or base radices, 64
bases 2 and 10, 137–40
Beit el Hekma, 44
binary number system, 24, 55–58, 68
binary-to-analog converter, reversible, constructing, 80–82
bits, 67
Book of the Dead, 12
brain hemispheres, 14
British Museum, 18, 19
bundle, 201
Byzantine Empire, 24
Byzantine era, 38

cancellation law, 155
Cantor Dust, 281–82
cardinal numbers, transfinite, 281
cardination, xii, 13, 153
Carmichael numbers, 151
Carmichael's variation on Euler's theorem, 150–51
carry, 72
Cartesians, 199
casting out nines, 113
ceiling, 230
chemistry, 4
cifra, 46
classes, 154

287

☥ Name Index ☥